Environmentality

New Ecologies for

the Twenty-First Century

Series Editors:

Arturo Escobar, University of North Carolina, Chapel Hill

Dianne Rocheleau, Clark University

Arun Agrawal

ENVIRONMENTALITY

Technologies of Government

and the Making of Subjects

Duke University Press Durham and London 2005

Designed by C. H. Westmoreland

Typeset in Cycles by Keystone

Typesetting, Inc.

Library of Congress Cataloging-

in-Publication Data appear on the

last printed page of this book.

Duke University Press gratefully

acknowledges the support of Rosina M.

Bierbaum, Dean of the School of Natural

Resources and Environment at the

University of Michigan, for providing

funds toward the production of this

book. Duke University Press is also

grateful to the University of Michigan's

Office of the Vice President for Research,

which provided additional funds toward

the book's production.

To my parents

Contents

About the Series

There is widespread agreement about the existence of a generalized ecological crisis in today's world. There is also a growing realization that the existing disciplines are not well equipped to account for this crisis, let alone furnish workable solutions; a broad consensus exists on the need for new models of thought, including more constructive engagement among the natural, social, and humanistic perspectives. At the same time, the proliferation of social movements that articulate their knowledge claims in cultural and ecological terms has become an undeniable social fact. This series is situated at the intersection of these two trends. We seek to join critical conversations in academic fields about nature, globalization, and culture with intellectual and political conversations in social movements and among other popular and expert groups about environment, place, and alternative socionatural orders. Our objective is to construct bridges among these theoretical and political developments in the disciplines and in nonacademic arenas and to create synergies for thinking anew about the real promise of emergent ecologies. We are interested in those works that enable us to envision instances of ecological viability as well as more lasting and just ways of being-in-place and being-in-networks with a diversity of humans and other living beings and nonliving artifacts.

New Ecologies for the Twenty-First Century aims at promoting a dialogue among those engaged in transforming our understanding and practice of the relation between nature and culture. This includes revisiting new fields (such as environmental history, historical ecology, ecological economics, or political ecology), tendencies (such as the application of theories of complexity to rethinking a range of questions, from evolution to ecosystems), and epistemological concerns (e.g., constructivists' sensitivity toward scientific analyses and scientists' openness to considering the immersion of material life in meaning-giving practices). We find this situation hopeful for a real dialogue among the natural, social, and human sciences. Similarly, the knowledge produced by social movements in their struggles is becoming essential for envisioning sustainability and conservation. We hope that these trends will become a point of convergence for forward-looking theory, policy, and practical action. We seek to provide a forum for authors and readers to widen—and perhaps reconstitute—the fields of theoretical inquiry, professional practice, and social struggles that characterize the environmental arena at present.

Preface and Acknowledgments

When I was conducting field research for an earlier book on migrant shepherds in Rajasthan in western India, the headman of one of the villages where the shepherds lived told me, "The grazing common (*oran*) in the village is in a bad shape. Animals have eaten all the grass, villagers have burned all the trees [as firewood]. . . . The government works so hard to improve our country's forests. Isn't it also our duty to contribute, even if it means some sacrifice?" These high-sounding sentiments seemed on the mark since reports about declining vegetation and advancing desert were current in the media. But I found it hard to accept his statement at face value. He was defending the village council's enclosure of the village common. The ostensible reason, he claimed, was environmental protection. But the enclosure had another, unacknowledged effect as well. It reduced the fodder available to shepherds in the village and forced many to migrate over longer distances. Because of their absence from the village, they became less likely to contest local elections or seek a share in the development funds that the village council disbursed. Landowners, of whom the headman was one, had initiated the move for enclosure and won this particular battle. The action his words defended were so clearly against the interests of the shepherds, and the words fitted so intimately with prevailing international and national narratives of environmental decline and loss, that I could not credit what he said. I found myself discounting his argument that a desire for environmental protection had played at least the main role in the enclosure of the common. Today, after completing this research on forest protection by villagers in Kumaon in North India, I wonder if I was too hasty.

My formal research in Kumaon began only in 1989 when I went there to find out more about local forms of control and regulation over forests. But I had visited earlier. My first visit, in the summer of 1985, revolved around an interest in Chipko, the internationally known movement to conserve the forests of western Himalaya. I met and spent a week each with two charismatic leaders of Chipko: Sundar Lal Bahuguna and Chandi Prasad Bhat. But these meetings did not provoke in me the necessary curiosity to study the movement or the region. The region and its people were fascinating, but the movement itself seemed too well known and the words of its leaders too well worn from recital to numerous disciples.

I returned to Kumaon in 1988 and again in 1989–90. This third visit occurred after conversations with Katar Singh and Vishwa Ballabh at the Institute of Rural Management in Anand, India. Their account of the forest councils in Kumaon, on which they had already written a paper, seemed more reflective of the daily struggles of villagers than the high-flying, sometimes contradictory rhetoric that circulates around the Chipko movement. I returned to Kumaon twice in the mid-1990s in an effort to understand better the government of forests. I visited nearly forty villages, examining local records, assessing the state of village forests, and interviewing hundreds of Kumaonis. This book is the result of that research and ensuing archival and analytical work.

The book is an investigation of environmental politics. It draws on theories of new institutionalism and insights from Foucault's poststructuralist work on power and subjectivity. My objective is to elucidate the shift in knowledges and forms of government that accompanied changes institutionalized by the colonial government and its successor administrators in Kumaon. I suggest that these seemingly different changes in knowledges, politics, institutional arrangements, and human subjectivities concerning the environment are of a piece and are best understood when considered together. The concomitant study of these changes helps extend contributions from three streams of interdisciplinary environmentalist scholarship—on common property, political ecology, and feminist environmentalism. I explore the potential of a new approach to the study of environmental relationships. I call this approach "environmentality" and show how it brings together a concern with power/knowledges, institutions, and subjectivities.

The introduction begins with a story about the transformation of a particular individual, Hukam Singh, into someone who came to care about the environment as a result of the institutionalized practices within which he was located. To illustrate how changes in Hukam Singh's understandings are more widely relevant to our views about changes in subjectivities, I describe similar shifts in the views of rural residents of Kumaon more broadly. I do so by discussing two incidents:

the first is the violent protests against environmental regulations in the early 1920s that resulted in hundreds of fires in forests in the Kumaon Himalaya in India, and the second is based on my fieldwork experiences in the 1990s, which show how Kumaonis in the same part of the Himalaya have now begun to protect forests. The book analyzes and explains how these transformations in politics, institutions, and subjectivities took place in the Kumaon Himalaya. The answer is relevant to the politics of community, decentralization, and subject formation in other parts of the world. Kumaon has the world's oldest surviving instance of formal state-community partnerships to govern forests. Explaining why Kumaon's residents began to protect their forests is important because local populations in more than fifty countries around the world are now engaged in similar efforts to protect environmental resources.

In the first part of the book, I look at how new technologies to govern the environment have emerged over the past 150 years in India (chapter 2) and Kumaon (chapter 3). I focus especially on statistics and numbers as forms of colonial knowledges that made new ways of governing forests possible. Changing technologies of government created forested environments as a domain fit for modern forms of regulation. But the attempt to govern forests through centralized instruments also provoked high levels of opposition from those who were already dependent on forests.

To tame the violent and fierce opposition of Kumaon's residents, the colonial state refined a new technology of government—decentralized regulatory rule—in the 1920s. Decentralization of environmental regulation involved changes in three sets of relationships that are the concern of the three chapters in the second part of the book. The first of these relationships concerns the ways in which states interact with localities. Decentralization of regulation produced what I call *governmentalized localities*. New centers of environmental decision making emerged in Kumaon after the 1920s. I examine them in chapter 4. Second, new regulatory bodies emerged to shape social environmental interactions in local communities. I call them *regulatory communities* and show in chapter 5 how their birth created new alliances and divisions among local residents and their representatives. Some local residents favored the institutionalized protection of forests through their village communities. Others were recalcitrant. I explain why. Finally, new technologies to govern forests were (and are) linked with the constitution of *environmental subjects*—people who have come to think and act in new ways in relation to the environment. Chapter 6 examines the conditions that have changed some residents of Kumaon, but not others, into environmental subjects. Widespread participation in self-regulation under perceptions of scarcity and need is the crucial mechanism that explains why many residents of Kumaon have become environmental subjects. Over the period considered (1850–2000), the joint changes in these three sets of relationships—between states and localities, community decision makers and common residents, and people and the environment—have led to the new technologies of government that the book explores and explains.

The last chapter of the book critically reviews the scholarship on common property, political ecology, and feminist environmentalism and shows some of the ways in which they can be combined so as to understand better the changes in environmental government as they occurred in Kumaon. It combines them in the approach I develop and call environmentality. Lest my review appear too critical, I should say that my initial interest in and exposure to environmental politics comes from my engagement with these literatures; I hope to talk about them as a sympathetic insider rather than an objective outsider.

This book has been long in the making, and every writer knows the loneliness of the enterprise. I am happy to report that loving friends were always at hand to relieve any rigors I encountered. Much of the book matured together with my friendships with Clark Gibson, Shubhra Gururani, Donald Moore, Jesse Ribot, James Scott, Ajay Skaria, and K. Sivaramakrishnan. They have taught me, as do all friends, as much as I was willing to learn. And they hid, with grace and forgiveness, any frustration at my inability to learn. I am especially thankful to Donald for his constant encouragement and unwavering faith in the project and to Jim for his willingness, always, to think with and against and for erring, if he does, unfailingly in favor of generosity. May he always do so.

A significant proportion of this book's arguments came to me while I was a postdoctoral fellow at the Workshop in Political Theory and Policy Analysis, working with Elinor Ostrom. There are few mentors who so brilliantly combine the personal with the professional, the family with the academy. No one does it better than Lin and Vincent. To Steve Sanderson I owe the debt of putting to paper my early reflections about community and conservation. The paper he invited me to write for the nascent Conservation and Development Forum at the University of Florida encouraged me to push further my burgeoning thoughts about how community works in environmental regulation.

Individual chapters have benefited from the careful reading and critical comments of many colleagues. I would like to thank Arjun Appadurai, Venkatesh Bala, Robert Bates, Amita Baviskar, Jacqueline Berman, Steve Day, Steve Dembner, Geoffrey Garrett, Michael Goldman, Franque Grimaud, Michael Hathaway, Margaret Levi, Margaret McKean, Vincent Ostrom, Nancy Peluso, Pauline Peters, Kimberly Pfeifer, Kent Redford, Robert Repetto, Suzana Sawyer, Leslie Thiele, and Mary Beth Wertime. I am sure I have missed others but know that some of these are not too displeased at my forgetfulness. After all, being acknowledged must make those being named party to arguments that they might only recognize as alien.

Sharad Tapasvi helped me at a crucial early juncture by tracking down many sources at the Uttar Pradesh State Archives in Lucknow, India. Other friends contributed enormously during the later stages of writing. Amy Poteete and Neera Singh suffered through the first draft of the manuscript and made numerous valuable suggestions that have clarified the language as well as the main points of discussion. Ashwini Chhatre, Tania Li, Victoria Murillo, and James Scott contributed encouraging and critical comments on the entire manuscript. I inflicted the book, with barely concealed haste, on my students at McGill University. Their reactions proved highly instructive. Several colleagues at Yale University persuaded me that they enjoyed reading parts of the manuscript. Through such subtle subterfuges did Kay Mansfield, Rogers Smith, Jessica Stites, and Eric Worby encourage me to continue work on the book. Kay now owes me a second bottle of single malt scotch.

Part of the book, mainly chapter 5, draws from previously published articles. I would like to acknowledge the following journals and publishers for their willingness to let me draw on work that was first published with them: *Comparative Political Studies, Development and Change, Journal of Agricultural and Environmental Ethics*, and *Journal of Asian Studies*. A different version of chapter 6 appeared in *Current Anthropology* in 2005 as "Environmentality: Community, Intimate Government, and the Making of Environmental Subjects in Kumaon India" 46 (2). Some initial thoughts are available in a succinct form in a short article I wrote, entitled "The Regulatory Community," for *Mountain Research and Development*. For their willingness to let me adapt sections of coauthored works for this book, I thank Gautam Yadama and Sanjeev Goyal. Both have taught me a lot more than an appreciation for statistics and formal modeling. I share with them a camaraderie that exists beyond the compass of this book's arguments.

I presented several chapters at professional meetings and as invited lectures. I would like to acknowledge comments and suggestions from members of audiences at the Seminar on Landed Property Rights, Social Science Research Council, Stowe, Vermont (1994); Indiana University (1997, 2003); Yale University (1997, 1999, 2000); Institute for Development Studies, Sussex (1998); American Political Science Association meetings (1998, 2000, 2001, 2002); University of California, Berkeley (1998); Emory University (1999); University of Minnesota (1999); University of Washington (1999); McGill University (1999, 2000, 2003); Harvard University (2000); Princeton University (2000); University of Michigan (2001, 2003); and Columbia University (2001). They showed remarkable patience and provided exemplary criticism and suggestions.

I completed the first draft of the manuscript at Yale University while on leave from teaching through a senior faculty fellowship. Revisions to the initial draft became possible with additional time that Ian Shapiro facilitated as chairman of the department: he reallocated my teaching responsibilities over different semesters. I owe William and Barbara Graham many thanks for making it possible for Rebecca and me to stay in Currier House at Harvard University. Laura Stephens generously offered me the use of her library office in Cambridge.

Grants from the Division of Sponsored Research at the University of Florida, the World Wildlife Fund, and the Social Science Research Council made possible the initial fieldwork. The writing of the book benefited from additional funds from the National Science Foundation (grant number SBR 9905443) and the MacArthur Foundation (grant number 0063798GSS). Funds from the DRDA and the SNRE at the University of Michigan assisted the publication of the book. For such expressions of confidence, I am grateful now and hope to be even more grateful in the future!

To Rebecca, who has taught me more about the ideas contained in this book than has anyone else, I owe what cannot be acknowledged. May we continue to learn together!

1

Introduction:

The Politics of Nature and

the Making of Environmental Subjects

To reflect upon history is also, inextricably, to reflect upon power.
—Guy Debord, *The Society of the Spectacle*

I

I first traveled to Kumaon in 1985 to learn more about how and under what conditions local residents protect their forests and the environment. At that time, I met a number of leaders of the Chipko movement —well known in India as a grassroots, collective effort to protect trees by means of direct social action.[1] But the meeting that left the most lasting impression was to occur in a small village by the name of Kotuli. Hukam Singh, a young resident of the village, told me that it was futile to try to save forests. Too many villagers cut too many trees. Too many others did not care. He himself was no exception. "What does it matter if all these trees are cut? There is always more forest," he said. He judged that only a few villagers were interested in what I was calling the environment. "Women are the worst. With a small hatchet, they can chop so many branches, you will not believe." He qualified himself somewhat: "Not because they want to, but they have to feed animals, get firewood to cook."

Hukam Singh's judgment is probably less important for what it says about processes of environmental conservation in Kotuli than for what it reflects of his own position. Talking with other people, I realized that the long periods of time Hukam Singh spent in the town of Almora prevented him from appreciating fully the efforts afoot to protect trees and forested environments. He was trying to get a job in the Almora district court and had stopped cultivating his agricultural holdings. But the village's forest council held meetings every other month and en- forced rules to regulate forest use. The 85 acres of village forest was more densely populated with trees and vegetation than several neigh-boring forests. The village forest guard often apprehended those cut-ting tree branches or grazing animals in the forest illegally. Most vil-lagers did not think of the forest as a freely available public good.

The reasons my conversations with Hukam Singh had a more lasting effect than those with well-known Chipko leaders were to become apparent during my return visits to Kotuli in fall 1990 and summer 1993. In the intervening years, Hukam Singh had left Almora, settled in Kotuli, and married Sailadevi from Gunth (a nearby village). He had started growing two crops on his plots of irrigated land and had bought several cattle. He had also become a member of Kotuli's forest council after one of his uncles, a council member, retired.

More surprisingly, Hukam Singh had become a convert to environ-mental conservation. Sitting on a woven cot, one sturdy leg tapping the ground impatiently, he explained one afternoon: "We protect our for-ests better than government can. We have to. Government employees don't really have any interest in forests. It is a job for them. For us, it is life." He went on: "Just think of all the things we get from forests—fodder, wood, furniture, food, manure, soil, water, clean air. If we don't safeguard the forest, who else will? Some of the people in the village are ignorant and so they don't look after the forest. But sooner or later, they will realize this is very important work. It is important for the country, and for our village."

These different justifications of his transformation into someone who cares about protecting trees are too resonant with prevailing rhet-orics around environmental conservation to sound original. But to dis-miss them because they are being repeated by many others would be to miss completely the enormously interesting, complex, and crucial, but understudied, relationship between changes in government and related shifts in environmental practices and beliefs. This book seeks to track such changes by examining emerging technologies of environmental government and their relationship with changes in human subjec-tivities.

Hukam's story mirrors the experiences of many in Kumaon. But equally there are others whose senses of the environment, relationships with environmental government, and actions in forested environments have changed relatively little, or may even have become more extrac-tive. Explaining why, when, how, and in what measure people come to develop an environmentally oriented subject position is the ultimate target of this book's arguments. These questions, provocative for both their practical import for conservation and their theoretical relevance to discussions of identity, require a historical examination of different technologies of government. New environmental subject positions emerge as a result of involvement in struggles over resources and in relation to new institutions and changing calculations of self-interest and notions of the self. These three conceptual elements—politics, institutions, and identities—are intimately linked. In exploring them together as constituent parts of a given technology of government, this book suggests that an exclusive focus on politics, institutions, or subjectivities likely leads to lopsided analyses of environmental politics and change.

The political, institutional, and identity-related struggles that I describe unfolded in the rich environmental history of Kumaon after the 1860s. Hukam's transformation in a sense constitutes a microcosmic window on changes in Kumaon. To explore and explain these changes, it will be useful to begin with two stories. One involves widespread protests against environmental policies in the early part of the twentieth century; the other, paradoxically, is about equally widespread involvement in environmental government that began around the 1930s and continues today. The shift that these stories represent can be understood by examining the emergence of new technologies of government that incorporated rural localities into a wider net of political relations, produced new forms of regulation in communities, and helped create new environmental subjectivities.

Massive forest fires, only some of them the usual summer fires, raged in Kumaon in the early part of the twentieth century. Between 1911 and 1916, the colonial state reclassified nearly 80 percent of Kumaon's forests into reserves. Villagers found that they had limited or no rights left in the reserves. In response, they set fires in the reserved forests in a vivid spectacle of challenge to new forms of government over nature.[2] Fires were especially widespread in 1916. Nearly 200,000 acres were burned in hundreds of separate incidents. As one observer noted: "An exceptionally dry state of the forests . . . and an outburst of incendiarism combined to create the worst record since fire protection was introduced" (Champion 1919: 353).[3]

Villagers set fires again and again in some places. In Airadeo, for example, fires burned for three days and two nights, and "new fires were started time after time, directly a counter-firing line was successfully completed" (Champion 1919: 354). In 1921, villagers set fire to even larger areas of forests, collectively protesting against the new regulations.[4] Forest and revenue department officials complained unremittingly about the difficulty of apprehending those who set fires. Burning beyond the power of the colonial state to control or extinguish, these fires would force a reconsideration of existing policy.

Official policy at the beginning of the twentieth century aimed to bring forests under centralized control. The colonial state in Kumaon Himalaya had insinuated itself deeply into processes of forest making. It had created and instituted entirely new procedures to control, manage, and exploit landscapes it deemed valuable.[5] The forest department had carried out surveys; demarcated different categories of forests; made working plans for planting, management, and rotational harvesting of trees; limited grazing by domestic animals; restricted collection of fodder and firewood; and introduced fire protection.[6] These measures were part of a new technology of government that had already greatly raised state revenues in many parts of South Asia, among them Bengal, Bombay, Burma, and Madras. In Kumaon, it also pushed villagers into violent protests that the colonial state had not anticipated (Agrawal 2001a).

The fires set by villagers are indicative of something remarkable if for a moment we suspend "our compulsive concerns with causes and consequences to empathize properly with the phenomenon under consideration" (Zolberg 1972: 186). They suggest that the appropriation of ever more land and ever stricter enforcement had overstepped tolerable bounds. In the initial years of its existence, after 1860, the forest department implemented new regulations but also tolerated a certain level of illegality. The department was unable to enforce its new regulations to the letter, and villagers stubbornly continued with their existing practices. Reclassification, further new regulations, and stricter implementation in the second decade of the twentieth century were an unprecedented intrusion into the villagers' daily lives that they could not endure.[7]

Villagers did not just protest collectively and visibly. In collusion that was largely implicit, even those villagers who did not actively participate in protests would not reveal the identity of violators of the law. Collusion went beyond the common hill resident. Village headmen, appointed by colonial administrators, also refused to cooperate with foresters. What is more, the instances of planned incendiarism were just the proverbial tip of a vast iceberg of illegality. In direct violation of the new rules, villagers grazed their animals, chopped and collected firewood, felled timber, and harvested fodder. They had always done so. But the new restrictions and enforcement had criminalized every-day behavior by making illegal a range of what might be called custom-ary uses of forests.[8] By simply continuing to do what they had always done, villagers committed acts that had become illegal.[9]

As the social, political, and economic costs of the new forest regula-tions mounted, the colonial state in Kumaon appointed a three-member committee to investigate villager protests. The Kumaon Forest Griev-ances Committee toured the entire region and interviewed nearly five thousand villagers. Afterward it recommended that the government of the United Provinces should permit villagers to take formal control over most of the forests that had been reclassified between 1911 and 1916. It also suggested that villagers should be permitted to govern their forests under a general set of framing guidelines. The colonial state accepted these recommendations. The consequences have endured.[10]

Figure 1 graphically depicts the information on forest-related crimi-nality for some of the early years of the twentieth century.[11] It shows the conspicuous increase in forest-related convictions and then their dra-matic and equally rapid fall. This decline in prosecutions and convic-tions signals the beginning of a profound transformation in the charac-ter of forest control in Kumaon, the institutionalization of regulation, and, relatedly, in environmental identities.[12] The reduction in forest-related "criminality" was accomplished through a new technology of government. The transformation continues today, fueled (literally) by the transfer of thousands of square kilometers of forested land to vil-lagers. Kumaonis have formed more than three thousand village-level forest councils (van panchayats) to govern their forests. Spread through-out the length and breadth of Kumaon, these organizations have now become the source of protection for nearly a quarter of its forests. The legal basis for their existence lies in the Forest Council Rules of 1931, which the colonial state created following the recommendations of the Kumaon Forest Grievances Committee.[13]

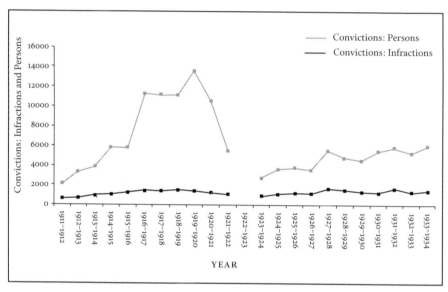

FIGURE 1: Convictions for infractions of forest laws in Kumaon, India, 1911–35.

Source: Annual reports of the forest administration in the United Provinces, 1912–36.

This book about environmental politics describes and analyzes how the government of environment has changed over the past 150 years in Kumaon and the relationship between changing technologies of government and the production of environmental identities. It examines the strategies of knowledge and power that created forested environments as a domain fit for modern government, focusing especially on the role of statistics and numbers in characterizing and reconfiguring forests (chapters 2 and 3). But technologies of government are not just about the formation of a new sphere—forested environments—in which power can be exercised. They are also about three other sets of relationships.[14]

They are about shifts in the relationship between states and localities. Such shifts produced what I call *governmentalized localities*. New centers of environmental decision making within localities emerged in Kumaon starting in the 1920s. Their interactions with the state have a considerably different tenor from the mainly antagonistic ones between the state and localities that existed earlier (chapter 4). A second part of new technologies of environmental government is the emergence of new regulatory spaces within localities where social interactions around the environment took form. I call these *regulatory communities*. Their birth meant new alliances and divisions among local residents and their representatives. Some local residents favor the institutionalized protection of forests that was being enacted in village communities. Others continue to be recalcitrant in the face of efforts to make the government of forests more efficient (chapter 5). Finally, new technologies to govern forests are also linked to the constitution of *environmental subjects*—people who have come to think and act in new ways in relation to the environmental domain being governed, forests. Of course, not all Kumaonis have become environmental subjects. I examine the reasons that account for the variable relationships between different Kumaonis and their environment as they see it (chapter 6). Over the period considered (1850–2000), the joint changes in these three sets of relationships constituted the new technologies of government that I seek to explore and explain.

The major concerns of this book are thus located on the shifting grounds of politics, institutions, and subjectivities that together characterize government in the sense of the "conduct of conduct" (Foucault [1978] 1991). Conduct of conduct can be inspired by many sources—agencies of the state, certainly, but also amorphous regulatory norms and institutions that affect the very thoughts and experiences of persons; authoritative figures, as within a community or family; or, as importantly, one's own self. To illustrate and elaborate how conduct might be shaped by some of these influences, I build on a number of writings in the field of environmental politics, especially by scholars writing about common property, political ecology, and feminist environmentalism (chapter 7).

My focus is on environmental government in Kumaon during the past century and a half. But the developments I analyze, especially those that occurred after the passage of the Forest Council Rules in 1931, presage processes that are now beginning to shape the politics of environmental policy in almost every developing country. New policies for the environment aim to decentralize government and secure the participation of local populations (Agrawal 2001b; FAO 1999). Policies aiming at greater decentralization and participation are about new technologies of government. To be successful, they must redefine political relations, reconfigure institutional arrangements, and transform environmental subjectivities. So, although the arguments in this book are advanced mainly as a way to understand developments in Kumaon, the discussion may usefully inform analysis of environmental government in other parts of the world as well. In this way, it is a useful lens to focus on swirling debates regarding public-private boundaries; the role of communities and states in environmental control; appropriate goals of environmental management; and discussions about resistance, domination, and subjectivity.

I propose *environmentality* as a useful name for the conceptual framework I use.[15] A union of *environment* and Foucauldian *governmentality*, the term stands for an approach to studying environmental politics that takes seriously the conceptual building blocks of power/knowledges, institutions, and subjectivities.[16] My analysis builds on existing writings by political ecologists, common property theorists, and environmental feminists. The arguments in this book illustrate the productive possibilities in the emergent interrelationships among these three concepts (see chapter 7).

The Forest Council Rules of 1931 have undergone several revisions as part of an effort to fine tune regulation. They continue to shape how forest councils protect forests.[17] They are the formal ground on which different agencies of the state relate to village forest councils. They also guide Kumaon's villagers in creating organizations to protect forests, designing rules to regulate actions, policing compliance with rules, apprehending those who do not comply, and meting out punishments to rebellious breakers. In all these ways, the rules prompt the councils to do the work of the forest department. The records of the forest councils greatly expand the realm of visibility for officials in the revenue and forest departments. Today Kumaonis control themselves and their forests far more systematically and carefully than the forest department could.[18]

My research on localized environmental regimes in Kumaon illustrates these inferences. In 1989, the first year of my field research in Kumaon, I organized a meeting in the district capital of Almora to discuss the problems council members faced. This meeting revealed many differences among villagers involved with the councils and village forests. There were differences between headmen and ordinary residents, upper and lower caste members, and men and women. Later I came to appreciate that other differences mattered even more in shaping environmental subjectivities: different types of participation in regulation, different forms of involvement in councils, and different levels of benefits from forests. I learned that by attending to practices of regulation more closely one may trace a more lived and living connection between subjects and power, environment and actions, and institutions and identities. Reading the politics of subject formation off the social categories of gender, class, occupation, and caste serves at best to ignore how power works to create the subjects who fill these categories.

More than forty council headmen, together with many forest users, attended the meeting in 1989.[19] Many of the headmen complained about village residents who illegally harvested fodder and fuelwood from council-managed forests. Equally pervasive were their complaints about officials in the forest and revenue departments. But these latter complaints were very different from what I expected. Instead of denouncing strict enforcement of forest laws, council headmen were censuring officials for their lack of enforcement. They said that government officials were too busy to help apprehend and punish villagers who had failed to follow forest protection rules.[20]

Villagers attending the meeting were subject to the rules that the forest councils had crafted. But, rather than questioning the legitimacy of enforcement rights vested in the councils, most affirmed the problems of monitoring and enforcement faced by the councils. They remarked on the scarcity of firewood and the need for alternative sources of fuel for cooking. They talked about the difficulties women faced in gathering adequate amounts of fodder from the forest. They described how onerous it would be to graze animals were they not to have access to fodder in the forest. And, to my surprise, many also mentioned that the environment was becoming more fragile and needed greater protection. The carping of the forest council headmen and the arguments of the villagers were borne out in the comprehensive Kumaon surveys and more detailed fieldwork I conducted in 1992 and 1993. I would begin to discern the reasons behind the seeming conflicts in these testimonies only with a better appreciation of the relationship between regulatory environmental practices and subject formation (chapter 6).

Between 1990 and 1993, I studied thirty-eight villages where forest councils regulated village forests.[21] New information underlined the earlier verbal testimony from council headmen. It showed that the statements of villagers about the need to protect the environment expressed in the 1989 meeting in Almora were shared widely. Certainly, some village residents complained about the existing forest management regime, and many admitted to breaking rules. But as I analyzed information in council records several patterns began to emerge. I could begin to answer questions about who broke rules and how frequently, who was apprehended by councils and guards, and which groups bore the burden of seemingly equitable enforcement of equal rules. It also became obvious that while villagers grumbled about enforcement they also accepted the need for protection.[22]

Many councils have created systems of monitoring that involve all village households. In other instances, resident families contribute to the salaries of guards appointed by councils. In most villages, even where the councils pay guards out of funds derived from the sale of forest products, villagers agree that their forest is in a precarious position. Infractions by villagers, it seems, vary in proportion to their expressed beliefs about environmental scarcities and the need to protect forests. The tenor of villagers' statements, in turn, seems to vary in relation to their involvement in environmental enforcement. Forest councils have created a number of monitoring and enforcement procedures that require different levels of participation by village residents. Villagers who participate more intensively in the enforcement of environmental protection also seem to care more about the environment. Contrary to my expectations, categories based on gender, wealth, income, and caste turned out to be less relevant as indicators of whether a particular person is likely to be interested in protecting village forests.

One must therefore be careful when interpreting differences between council officials and ordinary village users. True, there are some obvious distinctions between those who make and enforce the rules and those who use the resources and sometimes break the rules. But these differences do not mean that villagers generally consider locally enforced rules unfair or the efforts of the councils illegitimate. Conversations with villagers and forest council officials reveal some ambivalence about enforcement efforts but agreement that protection is necessary. Many, among them men and women, upper and lower caste members, and rich and poor households, emphasize the need to limit harvests in forests. They agree that forests today are in worse condition than they used to be and that institutionalized attempts to protect them are necessary and legitimate.

These views are in stark contrast to those prevailing at the beginning of the century when Kumaon's residents were setting fires in forests and refusing to inform on those whom the state considered criminals. They show that many residents of Kumaon have changed profoundly in their actions and views about forests. From being fire-wielding, state-defying, rebellious hill men, many of them have been transformed into members of a decentralized government-in-community inscribed on modern forests.[23] Following such shifts in sentiments about government requires careful investigation of knowledges and institutions, practices and politics, regulations and subject positions. After all, not all the hill men who had been "impatient of control," as the committee investigating their actions remarked (KFGC 1921), have come to participate in regulation or become environmental subjects.

II

The two environmental stories I have recounted—forest fires and the survey of Kumaon's residents—span a century. They are convenient windows through which to view the efforts of the Indian state to create new forms of environmental government. The first attempts to establish a forest bureaucracy can be dated to 1805, when the British East India Company established the first conservancy in Malabar (Stebbing 1922–26; see also the introduction to part 1, this volume). This early "forest conservancy" had little to do with the desire to conserve forests; rather, it was squarely implicated in the militaristic designs of an imperial power that needed timber in a transnational political economy.[24] Nonetheless, the 1805 conservancy was the forerunner of a mighty bureaucratic machine whose actions in Kumaon were directly responsible for an inferno of resentment. The fiery protests of the Kumaonis against forest conservancy destroyed property that the forest department had created in land, but they also prompted a new regime of regulation that ultimately penetrated far more intimately and precisely into daily acts of rural survival.

My survey of villagers' participation in and perceptions of forest regulations suggests that this infiltration of regulation into the unfolding of daily life was initiated as state policy but is not the coercive imposition that state actions are often taken to be. Nor are villager responses, when they violate existing local rules today, appropriately described by the term *resistance*. Indeed, analyses of villager interactions with procedures of rule over the long period under consideration are also only inadequately served by terms such as *negotiation* or *engagement*.[25] Such descriptions of interactions of villagers with regulatory measures hinge on prior conceptions about sovereign autonomous subjects that are simply impossible to identify in Kumaon and possibly elsewhere. They are also implausible in light of expressed statements from villagers about environment and government. Instead, regulations, and villagers' practices and words, seem to be part of a process that has reshaped people's understandings of forests and the basis of forest control itself.

The processes that reshaped forests and related institutions, practices, and subjectivities in Kumaon are examples of what might be called "governmentalization of the environment." Scholarship on government, with Foucault as the inspiration, has exploded since the 1990s (Baistow 1995; Barron 1996; Cruikshank 1994; Dean 1994a; Miller 1994; Miller and O'Leary 1987; O'Malley 1992; Procacci 1991). But only a few key interventions in the fields of environment and development have attempted to examine subject formation and its relationship with government (Li 2000; Moore 1998; Sivaramakrishnan 1999; Worby 2000). More generally, even in Foucault, and in much of the scholarship based on his suggestive arguments, there is only an exploration of the meanings of the term. There is little or no indication of *how* government shapes subjects or *how* one is to explain variations in transformations of subjects.[26]

In the nineteenth century, governmentalization of the environment was accomplished in India by the creation, activation, and execution of new procedures for surveying, demarcating, consolidating, protecting, planting, managing, harvesting, and marketing forests. Forest departments in different provinces of India created and instituted these procedures. By participating in these procedures, foresters' and state officials' basis of knowledge for viewing forests changed. Together with new ways to know forests, their subjective relationships with forests also underwent a significant shift. It would be a remarkable and possibly unprecedented triumph of cynicism if foresters in the nineteenth century were advocating exclusion from landscapes for economic gain but at the same time were completely aware that exclusion did not "improve" forests. Prior to the 1900s, communities and their members were directly affected by many of the new regulations that the colonial state implemented, and they were often cast in an antagonistic role in the processes of governmentalization that made forests.[27] That began to change in the early part of the twentieth century in some parts of India and especially since the late 1980s under the Joint Forest Management Program that is being implemented throughout the country.

At one level, new ways to govern forests were the result of changing perceptions about their potential uses, among them naval manufacture, sleeper ties for the railways, the production of turpentine, and of course revenue generation. Emerging demands because of greater commercialization, strategic imperial needs, and the consolidation of empire were crucial in shaping how state officials regarded forests. The exploration of potential teak supplies was born out of scarcities of timber for naval construction in England. The progressive appropriation of ever larger territories by the forest department could not have been accomplished if its revenues had not outstripped expenses consistently and ubiquitously in almost all Indian provinces. The defense of new procedures to govern forests was based on beliefs that private use could not be efficient owing to externalities that the public goods nature of forests made inevitable.[28] The proven utility of forests helped change beliefs about how they should be managed: governed as a resource rather than cleared to make way for agriculture.[29]

But at another level, new procedures and regulations based on statistical representations and numericized relationships also defined forests (chapter 2) and redefined legitimate ways to use them. They rendered some types of uses inappropriate and wasteful, illegal and ill-considered. They validated other types of uses on grounds of efficiency. They excluded some existing users. They favored others. They were mediating organizational forms in the widespread extension of new views about forests.

The application of new procedures to govern vegetation and land generated obstacles and resistance but also led to innovative ways of spreading official views of forests and their uses. Widespread opposition to forest conservancy in Kumaon signified that the government of forests required new partners in regulation. The birth of more than three thousand forest councils since the 1920s, one to govern almost every scattered plot of forest, was part of a process that officials debated from almost the earliest days of the forest department. Many in the forest and revenue departments in Kumaon talked about incorporating local populations in forest governance (Shrivastava 1996). Such discussions had also taken place in other parts of India. Consider Dietrich Brandis. Working as the conservator of forests in British Burma, he tried to persuade *toungya* cultivators—those who practiced "shifting agriculture"—to plant teak in the fields they cleared. In defense of his efforts to get villagers to sow teak, Brandis said as early as 1856, "if the people can be brought to do it, [it] is likely to become the most efficient mode of planting teak in this country" (quoted in Stebbing 1922–26: vol. 1, 378).[30] The forest councils of Kumaon constitute one of the earliest surviving attempts to give form to this vision of securing the participation of people in the making of forests. For that very reason, they form a crucial example through which we may gain an understanding of the long-term processes that emerge with decentralization of environmental government (chapter 3).

The Kumaon councils are analogous to the type of environmental government that more than fifty countries are trying to create today (FAO 1999). From Zimbabwe to the United States, from the Philippines to Cameroon, and from Mexico to Indonesia, national and state governments are striving to make rural populations accomplices in environmental and their own control.[31] The mechanism through which they seek to effect this transformation is the decentralization of environmental regulation to the locality, often through "community-based conservation." The success of decentralized efforts to govern the environment depends on the simultaneous implementation of three strategies: the creation of governmentalized localities that can undertake regulation in specified domains, the opening of territorial and administrative spaces in which new regulatory communities can function, and the production of environmental subjects whose thoughts and actions bear some reference to the environment.

The most visible of these three strategies is that which changes the relations of localities to central governments. It affects how localities or communities come into being and constitutes a redefinition of state-locality relations. This redefinition undermines the separation of localities from an administrative center and occurs in the context of contemporary forms of economic relations and political power. It leads to significant powers being vested in local officials but in a manner that allows superior officials to supervise more easily. In the Kumaoni context, thus, local communities can no longer be viewed as the autonomous village republics that Robert Wade (1994) describes in his fascinating account of regulation of irrigation in South India.[32] Nor can they be seen as the forms of social organization, predating capitalist transformations, about which Chatterjee (1983) talks in his analysis of the communal mode of power. As agents who are acting in the service of environmental regulation, the new communities are assuredly different from the vehicles of administration that the colonial state tried to craft by appointing paid headmen and revenue officials in Kumaon after the middle of the nineteenth century.[33] And ultimately these communities cannot be seen as instruments of despotic control. Governmentalized local communities, or, more briefly, "governmentalized localities," are part of a new regime of control that seeks to create fresh political-economic relationships between centers, localities, and subjects. They are knit together by the thread of state power. They are shaped anew by the soft hammer of self-regulation. They come to conform as a result of interventions that rely on knowledge about their internal dynamics (chapter 4).

The second aspect of the transformed relationship concerns how identifiable loci of power, decision making, and representation redefine the interactions between decision makers and common residents in communities. If governmentalization of localities denotes the emergence of tighter relationships between the state and the periphery, the development of closer links between local decision makers and residents is embodied in the dispersed "regulatory community." The dispersed loci of decision making emerge when communities are incorporated more thoroughly into wider circuits of political relationships. But the local actions of new decision makers are also an indication of the greater latitude that communities gain in crafting and extending formal regulatory rule over those who are resident within the territorially defined limits of the community. The power that decision makers such as general assemblies, forest councils, and forest guards exercise in territorialized communities is highly dependent on the people who are subject to their decisions. It is exercised in a circular manner. It depends for its effectiveness on a whole range of social, economic, and structural relationships between newly powerful decision makers and those affected by decisions. Formal processes shaping internal collective decisions help make communities agents in new regimes of regulation. Innovative regulations, often born within communities, extend the reach of power into the finest spaces of the social body. Social and institutional relationships within and between communities and their members come to be founded on the goal of a more strict and sustainable government of communal resources.[34] Thus, on the one hand there is the creation of new relations of regulations between the locality and the state; on the other hand, there occurs a transformation of the relationships between decision makers in localized communities and ordinary members of communities. What is being formed and refined is a legal mechanism of regulation through which actions in the forest can be calculated and manipulated and their legibility and visibility increased for outside observers. The counterpart of governmentalized localities is thus the "regulatory community."

But there is a third facet to the process, which is given the singular name of decentralization: the making of environmental subjects. Environmental subjects are those for whom the environment constitutes a critical domain of thought and action. This third facet is perhaps the most critical, ambiguous, and unpredictable and perhaps for these reasons the least well understood and investigated. It concerns how human understandings of and relations to forests change historically with the extension of centralized rule over forests and later with the emergence of the governmentalized localities and regulatory communities. These different institutional matrices of power produce different kinds of environmental subjects, as chapters 2 and 6 of this book explore. Rose (1999: 41) suggests: "There is a history to be written of the subjects of government." That many in Kumaon today think about their forests quite differently than they did in the early part of the twentieth century seems obvious. Evidence from the early twentieth century, it is true, is fragmentary and can be gleaned only through archival materials produced by state officials. Nonetheless, if we couple the available evidence with contemporary ethnographic and survey results, it is clear that the relationship of Kumaonis to forests and their ways of being in them have undergone (and are undergoing) momentous changes.[35]

These changes have occurred together with the creation of forest councils and more networked forms of power.[36] Power as it is practiced in the governmentalized locality and the regulatory community also environmentalizes subjects by changing how they view the environment and their place in it. Paul Veyne (1997) has argued that there is no universal subject over whom government acts. The emergence of environmental subjects in Kumaon is similarly about a process whereby local residents come to think about and define their actions, positively or negatively, in relation to the environment. And variations in subject positions are closely tied to practices and involvement in new regimes of monitoring, enforcement, and regulation.

Although this book suggests that effective decentralization processes are a combination of three different but connected changes, much analysis of indirect strategies of decentralized government focuses only on the first two relationships: governmentalization of subnational units of rule so as to extend state power, and the granting of limited autonomy to these units so that they can regulate the lives and practices of their members in specified domains.[37] Both relationships, for example, are the subject of Mamdani's (1996) examination of indirect rule and decentralized despotism in Africa.[38] The vast array of decentralized institutional structures around the environment that have come into being in the past two decades is also analyzed using a similar optic: what are the powers and responsibilities that the central state devolves to localized bodies; and what are the powers that these institutions, in turn, are able to deploy to regulate the lives of their members and the use of environmental resources?[39]

However, the ultimate success and effectiveness of these two strategies of decentralized regulatory rule depend also on shifts in the subjectivities of those undergoing regulation. Attempts to change how people act, when based solely on coercive threats in hierarchical organizations, are either formidably expensive or evidently impractical (Holmstrom 1982; Miller 1992). Indeed, analyses that depend only on causal arguments that are materialist and exclusively rationalist remain vulnerable to the charges Hegel leveled against Kant's moral philosophy: "excessive formalism, abstract universalism, impotence of the mere ought, and the terrorism of pure conviction" (see Calhoun 1993: 61). It is not surprising, therefore, that concerns about the relationship between changes in subjectivities and emerging social and political forms have elicited vast amounts of investigative energy from social theorists.[40] It is these same relationships between subjects and their environs about which Charles Taylor is talking when he asserts that "it took a long development of certain institutions and practices, of the rule of law . . . of habits of common deliberation, of common association . . . to produce the modern individual" (1955: 200).

In the case of the forest councils in Kumaon, historical and contemporary evidence suggests a significant link between changes in regulations, practices, and subjectivities. Indeed, access to material benefits from council-managed forests is at best a poor predictor of people's concerns about the environment or their willingness to protect it. The rational calculus of costs and benefits travels only a short distance in explaining the constitution of the subject itself.[41] On the other hand, the success of the forest councils in Kumaon also depends in significant measure on the production of environmental subjects: individuals who see the generalized need for environmental protection in some form and whose practices and words bear the mark of this acceptance, if not of personal conversion. To understand regulatory rule, therefore, it is necessary to examine how rule is experienced by those subjected to it. As decentralization of rule and local regulation become increasingly common in domains of political and social life such as the pursuit of development, infrastructure reform, welfare provision, health care, and education,[42] it becomes equally important to investigate how successful reforms in these domains hinge not just on institutional change but also on processes of subject formation.[43]

It is not necessary, of course, that all who are subject to regulations accept them as their own. Variations always mark the extent and pervasiveness of shifts in people's subjectivities. But the effective government of forests by the Kumaon forest councils depends in substantial measure on the willingness of at least a significant number of people to accept the rules that shape practices in the forests and to adopt as their own the processes of monitoring and enforcement. As more distant and scattered areas are influenced by the governmentalization of the environment, regulation of forests comes to depend on specific variations in rule making, monitoring, and enforcement strategies. Participation in these same institutional mechanisms is also at play in the redefinition of people's interests in forests and their subjectivities (see chapters 5 and 6).

III

Three important issues may be at stake as nation-states involve communities and local populations in the management of environmental resources.[44] Many countries in the developing world face a fiscal crunch: they need to reduce costs and become more efficient. At the same time, international donors have begun to make funds available for the design of new mechanisms of cooperation and new technologies of government that would make partners out of local actors. Finally, states have also come to recognize that the protection of environmental resources does not necessarily require exchange-based or coercive solutions. Indeed, coercion is often ineffective, and attempts to craft new modalities of exchange are often defeated by the public goods nature of environmental resources. In consequence, new programs of environmental governance in which communities play a prominent role have become widespread (Agrawal 1997). But their effectiveness and relationship to larger social processes remain unclear.

The ensuing chapters of this book attempt precisely to address this gap. The two-hundred-year history of changing forms of forests in India and the history of forest councils of Kumaon since the 1920s constitute a critical case study through which to understand changes in environmental politics. To examine this case effectively, I have divided the book into two parts. The two chapters in part 1 focus on the strategies of power/knowledge through which forests came into being in India and Kumaon. Chapter 2 shows how throughout the nineteenth century and the early years of the twentieth employees of the colonial state experimented with different procedures and mechanisms to gain greater and more efficient control over desired harvests from forests. New forms of representation of forests were based on the desire of colonial administrators for systematization, uniformity, and predictability in processes of control and exploitation (see Scott 1998). The period witnessed the creation of masses of numerical data on forests, which formed the basis of administrative and political claims to greater territorial control. By the end of the century, the forest department had asserted its right to administer close to a quarter of India's territory, nearly 250,000 square miles of land. New administrative and representational innovations brought into being what we today understand as Indian forests.

The forest department's efforts to implement uniform plans that would simplify and smooth the streams of revenue from its forests confronted enormous practical difficulties. Local exigencies and administrative rivalries reshaped the nature of environmental interventions and forced departmental officials to accommodate both human and nonhuman resistance. The development of new technologies, the emergence of new representations, and the specific uses to which forests could be put changed, as did the procedures for making and managing them. State objectives always demanded some level of social exclusion, often undertaken in the name of systematic government. Chapter 3 provides a careful empirical examination, based on evidence from Kumaon, of the initial processes through which the colonial forest department produced and governed landscapes. The discussion helps situate the responses of excluded populations and the subsequent dispersal of processes of regulation.

Of the different regional efforts in India to incorporate forests into circuits of imperial rule, the ones in Kumaon were among the most important, especially in terms of contributions to state revenues.[45] The history of Kumaon's forests throughout most of the nineteenth century is one of increasing insistence by forestry officials that forests are state property, that their protection is necessary for environmental health, and that the forest department is the best agency to institute such protection. The new regulatory regimes of forest control were based on ever greater usurpation of expert authority, claims to scientific knowledge, and the launching of what many scholars have termed "scientific forestry." They became central to appropriate forest management and indeed to the production of forests themselves. In the process, they effectively promoted the interests of the forest department in its intrastate institutional rivalries with the revenue department. These rivalries bore special fruit when they resonated with the local resistances that emerged as an initial response to procedures of control.[46]

The second part of the book takes a more deliberate view of environmental politics by examining changes in Kumaon more thoroughly. Its three chapters analyze the emergence of a new technology of government of the environment. They focus, in turn, on the creation of governmentalized localities, institutional politics within regulatory communities, and the making of environmental subjects. In the process, they illustrate how different aspects of environmental government—politics, institutions, and subjectivities—come together in the conceptual framework I call environmentality.

The costs of centralized bureaucratic control usually fell on shifting cultivators, graziers, villagers, and local merchants, whose subsistence practices and efforts to extract commercial gain from forests were sharply curtailed when the colonial state asserted the primacy of its own claims against existing patterns of access and use. Villagers' protests moved the rhetoric of forest management in a direction in which forest department officials slowly categorized local residents as uninformed and junior but potentially useful partners in managing forests for profit and local livelihood. The creation of village-level forest councils in Kumaon in the 1920s and 1930s generated perhaps the oldest surviving example of formal efforts to manage forests through the joint authority of the state and the community. Chapter 4 shows how the birth of forest councils transformed the character of the relationship between the state and the community. Instead of treating the community as the refuge of insubordinate forest users, forest and revenue department officials found it more convenient to consider it as a locus of regulatory authority. Treated thus, the social body of the community could become the context for the construction of new channels for the flow of power.

The local exercise of power has also resulted in a different basis for the involvement of Kumaonis in the governing of their forests. From being situated primarily as victims and opponents of control, many of them have become active participants in processes of environmental management and government. Many also participate in the selection of representatives entrusted with the exercise of control. Still others play an active role in the enforcement of state-sponsored as well as locally crafted rules for managing forests.

I use information from several sources to show the relative importance of regulation and enforcement in the conservation of forests in Kumaon. Chapter 5 contains detailed information on thirty-eight villages and their efforts to monitor environmental practices. It documents how the relationships between communities and their members have changed since the creation of the Forest Council Rules in 1931. Community leaders now enforce regulations to conserve and govern the forest rather than leading local residents into protests against regulation. New rules and procedures govern the environmental practices of villagers in forests. Fresh regulations change the calculus of interest on the part of members and lead many members of the community to become active participants in processes of regulatory control.

Just because villagers come to participate more intimately in environmental regulation, or at least become more environmentally concerned, does not mean that the effects of regulation are equally felt by new environmental subjects. In fact, variations in the way households are situated in multiple fields of power critically influence who bears the brunt of regulation. Allocation regimes of the forest councils and gender- and caste-related variations in sanctions on community members combine to produce substantial differences in the experience of regulation (Agrawal 2001a). The greatest adverse impact of enforcement is borne by the most marginal groups within villages.

The ubiquity of villager interactions in forests implies the necessity for a concomitant ubiquity of enforcement. An examination of the social basis of enforcement and regulation in chapter 6 shows that regulatory mechanisms that elicit widespread participation from a significant proportion of villagers are also effective means of transforming subject positions. Participation in regulation is not only necessary to generate the surplus that underwrites the processes of control, but it is also important in generating the concern for conservation that renders environmental protection a moral act.

Processes around the regulation of forests are thus the links that join the political to the perceptual, the managerial to the mental. By delegating regulatory authority to a set of representatives, and by direct involvement in administering, managing, controlling, and restraining collective behavior, rural residents come to construct their own stories of environmental decline, forest scarcity, and human threats to continued resource use. Regimes of regulation and enforcement transmit information about the nature of incentives users face. But involvement in the creation of new incentives and social practices related to an institutionalized structure of incentives also transform what users think and say about the environment. Regulation is not just about restraining a group of people who might break the rules. Much more importantly in Kumaon, and for a much larger group of people, regulation is the source of the awareness and recognition of the fragile re-

sources on which livelihoods depend and the context in which practices unfold (Bourdieu 1977; Calhoun 1993). The varying forms of regulation and the practices through which people get involved in them
are the basis for making environmental subjects.

The empirical study of the production of forests, communities, and
subjects in Kumaon prepares the ground for a more general analysis of
environmental politics, especially as it has evolved since the 1970s.
Chapter 7 elaborates the framework of analysis I call environmentality
by drawing on the strengths of three important literatures on the environment: political ecology, common property, and feminist environmentalism. These interdisciplinary writings have been instrumental in
exposing the many problems in centralized and exclusionist conservation strategies that were favored by most national governments until
the 1970s. States considered central interventions necessary because of
the public goods nature of the environment and were often impelled by
the presumption that poor indigenous groups or communities could
not act in their own long-term interest. The study of environmental
politics prior to the 1970s similarly ignored local users, communities,
indigenous peoples, and other marginalized groups except in viewing
them as obstacles to environmental conservation.[47] Environmental politics often meant a debate between advocates of privatization and the
devotees of centralized control.[48]

With the shift in conservation policies toward communities, a different but equally specific and limiting meaning of politics and environment has been consolidated. Recent analyses in the field of environmental studies, when they consider politics, usually treat the
environment as yet another arena in which traditional conflicts such as
those between elite and poor, state and community, indigenous and
outsider, or men and women unfold. They fail to take the environment
seriously. In arguing for a more serious consideration of the environment, I do not mean to suggest that it stands outside its human understandings and social constructions. Rather, I am advocating for a position that takes as its concern the experiences of environmental changes
and politics among those who are the subjects of scholarly investigations. To ignore changes in environmental subjectivities is to miss an
entire domain of politics and practice implicated in the making of
environmental subjects. At its core, environmentality is about the si-

multaneous redefinition of the environment and the subject as such redefinition is accomplished through the means of political economy. In this sense, it refers to the concurrent processes of regulation and subject making that underpin all efforts to institute new technologies of government.[49]

PART I

overleaf: Kumaon's middle Himalayas
Photo by Arun Agrawal

I

Power/Knowledge and

the Creation of Forests

In 1805, the Court of Directors of the British East India Company sent
an urgent dispatch from London to the Bombay government in India. It
wanted to know if teak from Malabar in western India could be sup-
plied for the King's Navy. The growing deficiency of oak in England
over the last three centuries had greatly increased the commercial and
strategic value of timber in the colonies.[1] The inquiry by the directors
led the Bombay government, in a somewhat bureaucratic fashion, to
appoint a Commission of Survey. The commission's mandate was to
assess Malabar forests and their ownership.[2] The commission reported
that the capacity of the forests had been overestimated, that the more
accessible forests had been almost depleted, and that costly road construction would be necessary to exploit the more distant teak-bearing
areas (Stebbing 1922–26: 64).[3]

Alarmed by the commission's report, the British East India Company
issued a general proclamation prohibiting the unauthorized felling of
teak.[4] The company also asserted its exclusive right to royalties that
were earlier claimed by indigenous rulers (Ribbentrop 1899: 62–64).
With an eye on regular timber supplies for the navy, and under pressure
from the home government, Captain Watson of the East India Company Police Service was appointed the first conservator of forests in
India in 1806.[5] Watson took the duty of supplying teak more seriously
than that of ascertaining whether it was harvested legitimately from
unclaimed lands or forcibly from private lands. He soon established the
company's monopoly over teak throughout Malabar and Travancore.
Watson's actions naturally alienated landowners. They even turned
timber merchants against the government.[6] Stebbing, usually a staunch
defender of imperial conservationist policies, is unequivocal in his cen-
sure: "The new regime was far too drastic to be continued as a method
of permanent administration. [Even] the privilege of cutting fuel for
private use . . . was invaded and prohibited, a short-sighted step of
amazing folly. . . . By 1823, the growing discontent of the forest proprie-

tors and timber merchants chafing under the restrictions of the timber monopoly, and the outcry of the peasants indignant at the fuel-cutting restrictions, came to a head. . . . [The] Conservatorship . . . was abolished" (1922: 71).[7] Systematic and widespread state efforts to intervene in how Indian forests were to be used, managed, regulated, governed, and viewed would begin again only in the 1850s.

This aborted conservatorship marks the first major effort by the British to initiate the government of forests on the Indian subcontinent. It foreshadows the unprecedented transformation of landscapes and subjectivities that was to occur as the British consolidated regulatory control over Indian timber.[8] Initial colonial efforts to extend control can be seen as part of larger imperial rivalries (Albion 1926; Bamford 1956) because forests were crucial to naval superiority. Indeed, they were the cornerstone on which depended "virtually all aspects of material culture" (Lindqvist 1990: 301).[9]

Although this first forest conservancy turned out to be different in many ways from more widespread efforts after the 1860s to assume control over India's forested lands, it prefigures them in four important ways. First, it points to the keen interest of the British in establishing control over vast areas of forests because of the commercial and strategic value of specific timber species. Second, it shows how burdensome such control would prove to be for many existing users and managers. Forests all over India had many and competing uses, and colonial control often required exclusionary practices. Third, it hints at the debates over different forms of private versus public control that framed the exploitation of forests for many decades to come. Finally, it indicates the willingness of the colonial state to change, and sometimes abandon, its preferred strategies of control when it confronted conflicts.

But there are also serious differences between the first attempt at forest exploitation in the early 1800s and the later, more considered, attempts through the establishment of forest departments. One dimension of difference lies in the later deployment of systematic organizational mechanisms founded on new forms of knowledge about trees and landscapes. The first conservancy focused on the extraction of teak and treated trees as commodities to be mined for profit. In this, it mimicked many indigenous interventions that aimed to extract timber for shipbuilding or other specific purposes. In contrast, forest departments founded from the 1860s onward developed and brought together new technologies of government that had as their underlying assumption a view of Indian forests as exhaustible resources with competing uses. Exhaustibility and competing uses necessitated systematic protection, regulation, and improvement, which varied with soil, climate, rainfall, vegetation mixes, tree density, and the effects of these on growth rates of timber. Emergent forms of statistical knowledge and the development of numericized relationships between land, trees, care, and profit gave later interventions a more refined guise. Statistics allowed forests to be apprehended summarily and unambiguously.

The two chapters in part 1 of the book examine the forms of rule that helped institutionalize new strategies of knowledge and power and some of the difficulties that such strategies encountered. Chapter 2 focuses especially on the role of statistics and numbers. In existing writings on Indian forests as a domain of government, questions of quantification and statistics have typically found little attention. Instead, scholars have mainly considered two issues. The first concerns the nature of the rupture that colonial practices inaugurated in Indian environmental history, especially in relation to forests. Whether colonial interventions marked an unprecedented double exploitation—of forested landscapes and peoples who depended on forests—is the key issue on which much of the debate turns.[10] A second question concerns the relationship of the present to the past and focuses attention on the significance of recent trends toward decentralization in Indian forestry. These trends are embodied most explicitly in what the Indian government has termed the Joint Forest Management Program.[11] In this instance, the issue turns on the extent to which recent shifts represent a radical break with a century and a half of paternalistic, top-down control that the forest department exercised over India's forest resources.

Both of these questions about historical understandings and the contemporary status of forests are mainly concerned with the political economy of forests and environmental changes. Current writings on these questions tend to ignore the representational aspects of Indian forests. As a result, they do not adequately explore a major colonial innovation in the government of forests: the use of statistics. In viewing colonial forestry as constituting a break between a harmonious golden past and a socially marginalized rural population, many scholars use an analytical framework that sets aside questions of another kind: how did colonial rule come to consolidate a certain view of forests that became hegemonic among colonized subjects?[12] By ignoring the continuities between the past and the present insofar as statistics and numbers are concerned, analysis risks conflating political-economic with representational regimes. Even if current policy changes toward greater decentralization mark a break from the technologies of government that centralized colonial rule was all about, the representational regimes around forests that emerged in the nineteenth century continue to this day. It is impossible today to talk about forests, for example, without reference to their area, functions, density, and ecosystemic characteristics. That we can talk about forests by referring to these features only became possible after the representational innovations of the mid-nineteenth century (see also Sivaramakrishnan 1999).

These representational innovations included new procedures for measuring, aggregating, differentiating, and analyzing. The specific features of vegetation that should receive the greatest attention and refinement in these new procedures were a matter of contestation but were ultimately determined by recourse to those great devices of commensurability: prices and profits. Conversion into monetary value ensured that the worth of different procedures could be objectively assigned. Numbers were obviously crucial both to measure and to establish equivalence. When applied to trees, the innovations of measurement, aggregation, differentiation, and analysis yielded relationships that described the effects of human interventions in the landscape. Thus, new forms of representation facilitated fresh strategies of intervention (without necessarily precluding earlier forms of exclusion). Among the new interventions we can count surveys and demarcation, thinning and clearing, plantations and working plans. There were many regional variations in the practice of these interventions, a result of perceived differences in the value, accessibility, and ease of exploitation of available timber. And these nineteenth-century forms of representing forests—in terms of numbers and statistics—have been bequeathed to contemporary technologies of government that involve communities and local populations as partners of state officials.

The consolidation of a new representational regime and associated changes in how forests were to be governed should not be interpreted as a totalizing process. Chapter 3 focuses on Kumaon to track responses to the new policies. Whereas chapter 2 examines the spread of statistics and the colonization of forestry professionals by numbers throughout

India—and hence tracks the continuities in this process—the subsequent chapter explores the divisions among different departments of the state that the new technologies for governing forests prompted. Statistics-based claims rose to prominence and became the means through which the state could appropriate ever vaster areas of land in Kumaon. These claims made it possible for the forest department to craft new legal restraints on Kumaon's residents, reclassify a vast proportion of Kumaon's land as forests, and justify its actions using the idiom of long-term sustainability.

The reclassification and appropriation of vast areas of land in Kumaon generated intense tussles within the state. Many officials favored the continuation of existing centralizing policies. Others supported decentralization of authority and the deployment of a new technology of government for the environment. The proponents of the first view tended to hold important positions in the forest department, and those holding the latter were mostly in the revenue department (Shrivastava 1996). Even those who supported decentralized government of forests did so on the grounds that it would yield more complete regulation of the environment by enmeshing Kumaonis more closely in the process of government and by making them accomplices in the project of regulatory rule. The forces and actors that laid the foundations for a decentralized technology of government are the concern of chapter 3.

2

Forests of Statistics:

Colonial Environmental Knowledges

A handful of mere statistics of the most routine, humdrum sort can sketch a picture of the basic characteristics of the Indonesian archipelago as a human habitat with more immediacy than pages of vivid prose about steaming volcanoes, serpentine river basins, and still, dark jungles.—C. Geertz, *Agricultural Involution*

The Indian Legislature, when preparing a forest law for India [in 1878], decided not to attempt a definition of forests, but merely to provide that the Government may declare certain lands to be forests and thus bring them under the operation of the Indian Forest Law.
—W. Schlich, *Manual of Forestry*

Beginning in the middle of the nineteenth century, Indian foresters came to rely increasingly on numbers and statistics to represent the land and vegetation they controlled. Numbers became crucial for recording and documenting forests. Calculations and mathematical relationships emerged as the basis of arguments about specific techniques for shaping forests. Statistics came to occupy a privileged place in the advocacy of different institutional and legal structures through which to govern them. Numbers, calculation, and statistics thus became a basic part of technologies of government over nearly a quarter of colonial India's territory by the end of the nineteenth century.[1] Not only did they help represent land and vegetation, but they also helped constitute it. What Churchill said about lived space, "first we shape our buildings and then they shape us" (cited in Alonso and Starr 1987: 3), can be said to have become equally true of the relationship between forests and statistics.

This chapter examines how measurements, numbers, and statistics—particular forms of power/knowledge—helped make India's forests. The objective is twofold: to describe the relationship between statistics and forests and the implications of this relationship, and to challenge our conception of forests as self-evident, material entities. My question is not whether forests existed prior to British rule, but rather how the idea of forests and perceptions about their materiality came to depend on the use of statistics. After the late-nineteenth-century, quantification became the preferred means of representing forests as well as advancing and defending claims over them. An exploration of the specific human interventions that helped determine the appearance, images, and referents of forests raises important questions about contingency, nominalism, and stability in the relationship between representations and objects (Hacking 1999: 63–92).[2] I address these issues at the end of this chapter.

My analysis takes as its point of departure much recent work on the origins of environmental transformations in South Asia and the political-economic impact of colonial forestry policies.[3] Although studies undertaken since the mid-1980s have ably debated political-economic changes in India's environmental regimes, they have been less concerned with tracing the equally elemental changes that colonial rule introduced in the representational economy of forests (cf. Sivaramakrishnan 1995). The distinctive and enduring role of statistics and numbers in the production of Indian forests has found only limited attention in general.

But the role of statistics has undoubtedly been crucial. Writing about its use as a means of understanding contemporary life, Raymond Williams observed that the society emerging out of the Industrial Revolution would literally have been unknowable in the absence of statistical theory and data (cited in Asad 1994: 67–68). Indian forests would not only have been unknowable without statistical representations, but such representations helped constitute the very category of Indian forests. Their extent, value, role, importance, and place in the national economy; demands for their protection and management; and concerns about the effects of human interventions can be appreciated only inadequately in the absence of numbers and the use of numbers to represent social facts.

To examine the special role of numbers and statistics in the making of forests, I draw particularly on historians of statistical representation.[4] I use their arguments to prepare the ground for examining the crucial role of quantification, statistical abstraction, and numericized relationships—all introduced in the late nineteenth century in relation to colonial Indian land and vegetation—in the making of forests. In talking about statistical representation, I have in mind the specific, commonsensical meaning of *statistics* as a set of methods for treating quantitative data and the data themselves and a genre of writing that employs numbers to describe entities and aggregates. In this sense, forests are an example par excellence of territorialized entities summarily and reductively represented by specific figures—area of land, number of species, volume of product, or density of vegetation cover—and relationships between age and girth of trees, girth and volume, land area and yield, or soil quality and wood volume increment.

By "making of forests," a phrase I use equivalently with "construction of forests," I want to stress that, like all representations, those relying on numbers do not merely reflect their social ground but supplement it and contribute to it as a result of the politics involved in selecting and highlighting specific attributes of an entity (Alonso and Starr 1987; Scott 1998). As representations gain credence, they reshape and substitute for the phenomena they purport only to describe. The making of forests in India thus involved a double erasure. The first was an erasure of earlier ideas about forests and their displacement by statistically oriented views, and the second was an erasure of the referents of those earlier ideas when statistical ideals worked recursively to reshape referents.

The ensuing argument can be briefly anticipated. Although the major legislation that asserted colonial control over Indian forests was not in place until the 1870s, administrators, natural historians, and botanists in different provinces had already begun to use the elements comprising a new political-economic regime for forests by the 1850s. Dietrich Brandis brought together most of these elements for the first time in Burma in the mid-1850s. He also played a crucial role in their territorial diffusion and institutional consolidation after he was appointed India's first inspector general of forests in 1862. Around the same time, the use of statistics and an "avalanche of numbers" (Hacking [1981] 1991) regarding various aspects of forest growth and management came radically to reconstitute views about forests. A whole series of practices related to naming, classification, counting, measuring, and valuing affected sense making around forests. My general argument is that the new representational and political-economic regimes that emerged around roughly the same period in the mid–nineteenth century were distinct but part of a single "colonial project" (Thomas 1994) of rule over the environment. It is in the conjunction of statistical thinking and biocommercial concerns that the novelty of colonial interventions in the environmental domain needs to be situated.

Investigating this conjunction is altogether a more complicated affair than is suggested by readings that identify the use of statistics and numbers with domination and control (Miller and O'Leary 1987; Pasquino 1991; cf. Rose 1999). For example, recent transformations in the allocational policies surrounding forests in India bring communities and local populations into a partnership with the state, but they continue to depend on representational mechanisms very similar to the more extractive policies that British rule often authorized. Advocates of decentralized as well as centralized government of forests thus wield statistical arguments in the defense of their goals. It would seem that statistical facts can be yoked to multiple roles. But it seems equally evident that once precise, statistical, generalizing arguments are invoked in the service of policy it is difficult to counter them with vague, descriptive, anecdotal evidence. It is in this characteristic of statistical representations—their capacity to displace nonnumericized arguments and advocacy—that their colonizing effects are to be found.

Debating the Environment

From almost the very birth of environmental writings, the major question confronting scholars has been about the degree to which colonial rule transformed how humans exploited natural resources. Ramachandra Guha refers to this question as leading to the birth of the "Great 'Ecology and Colonialism' Debate." He suggests that British rule introduced "rapid, widespread, and in some respects irreversible changes" (2000: 215). Colonial policies formed an "ecological watershed" (Gadgil and Guha 1992) because they were "socially unjust, ecologically insensitive, and legally without basis in past practice" (Guha 2000: 216). Resistance to British forestry policies, in consequence, was widespread throughout the subcontinent and even outside India.

The work of Richard Grove and some of his colleagues stands in contrast. Pushing further the arguments in Grove's (1995) authoritative history of colonial environmental ideas, Grove, Damodaran, and Sangwan (1998) suggest that the major outlines of British conservationist arguments had already been fleshed out in the early nineteenth century, a period Guha ignores in his work. They argue as well that conservationist sentiments were embodied organizationally in the forestry departments founded by colonial rulers throughout the empire. Skeptical about the survival prospects of customary land and forest use regimes that Guha defends, they use evidence from other British colonies that did not have a forest department to argue that "without exclusionist forest reserve legislation, most surviving forms of 'common property management' would have faded away like snow on a summer day" (14).

Echoing Theodore Porter's remarks about the history of quantification, it would not be far wrong to say that in these polarized positions one detects the arguments of partisans who for the moment have forgotten the value of nuance (1996: 6). However, painting in such bold strokes has had some positive impact as well. One salutary effect has been the explosion of work on the ecological legacy of colonialism.[5] There is also now greater nuance in how scholars of the environment address questions around gender, social mobilization, and the relationship between the agrarian world and the environment.[6] And, finally, new research has emphasized the value of indigenous science and its contributions to taxonomy, silviculture, and medicine.[7]

But the very intensity of the debate may have helped channel research into predictable directions related to the political economy of resource regimes. Little of the existing work on forests (or other resources) investigates the representational economy of resource control. Investigating the role of numbers and statistics in transforming the management and production of vegetation for the needs of colonial rule is thus a potentially fruitful arena of inquiry since government depended on these novel interventions to define and refine the role of timber and trees in the social landscape and the larger economy. Some suggestive leads in this direction can be found in the works of Demeritt (2001), Scott (1998), and Sivaramakrishnan (1999). This chapter takes the arguments they present further by focusing on the representational economy of forests and the environment rather than their political economy—a ground well covered.

The use of numbers to represent social facts exploded around the mid-nineteenth century with the invention of statistics. Since forestry departments were established in most imperial possessions only after the second half of the nineteenth century, it is not surprising that their reliance on statistics increased over time. In the eighteenth century and the early nineteenth, many natural historians, medical professionals, soldiers, administrators, and travelers produced accounts of trees and forests in India. These writings were mostly detailed narrative descriptions. Typically, they were founded on individual experiences. They scarcely employed numbers or quantification. But by the end of the nineteenth century, forests in official writings and memoranda had a very different character. Numerical representations, purporting to describe what was happening to India's vegetation and timber wealth, had been important first in advancing the rationale for an independent forestry establishment. But foresters began to use statistics as their preferred means of conveying meaning after the 1860s and 1870s. An entire organizational apparatus began to collect data actively and constantly. Standardized techniques of data collection increasingly became the basis on which all foresters were trained. Numbers came to mold the way the world of trees and vegetation would be regarded and understood.

Categories preceded numbers. Classification of the landscape into different zones of vegetation types and identification of constituent species within each zone were foundational to the assignment of different kinds of numbers to represent each category summarily. But numericized vegetation—number of trees, volume of wood, area of forest, revenue per unit of volume—helped foresters and other administrators grasp an entire region, indeed the entire imperial forested estate, without ambiguity. By estimating statistically the relationships among constituent units within their estate, foresters advanced proposals about its management. The concreteness of numbers was far superior to the rough and ambiguous adjectives and superlatives that had earlier described India's vegetation and trees.

Beginning in the last decades of the nineteenth century, foresters used numbers to contest each other's views about the state of vegetation, trends in the health of forested estates, and policies to shape the constitution of vegetation and its productivity. But the perceived concreteness of numbers was not just a faithful representation of an underlying uncontested reality. "The questions asked (and not asked), categories employed, statistical methods used, and tabulations published" depended on political and economic choices about what to measure, how to measure it, how often to measure, and how to present and interpret the results (Alonso and Starr 1987: 2–3). Such choices inevitably continue to shape how numbers represent and numerical representations recursively affect actions concerning "that which is represented."

New Classifications and the
Production of Forested Landscapes

Stebbing's magisterial three-volume study of forests in British India was written after more than a century of colonial explorations and annexations in some of the most remote regions of India. It begins with a familiar, authoritative, and precise classification (1922–26: 41–67).[8] Forests fall into six zones: evergreen, deciduous, dry, alpine, riparian, and tidal. These zones are further subdivided into geographical regions: "The evergreen zone may be subdivided into four distinct geographical regions" and so on (41). For each region, Stebbing reports the boundaries that separate it from other regions. For example, the Sub-Himalayan region "covers the belt of low country bordering on the spurs of the eastern Sub-Himalayan range, entering deep into their valleys and covering the slopes of the lower spurs" (43). Within each region, he lists the names of major families of vegetation and commercially valuable or dominant trees. For example: "The number of species composing the forest is very great and the trees individually attain a great size, of which the most important are the following: *Schima wallichii, Terminalia tomentos* and *Myriocarpa; Artocarpus chaplasha; Cinnamomum glanduliferum, Echinocarpus sterculiaceus, Bombax malabaricum, Dillenia indica, Eugenia formosa* and *Pterospermum acerifolium*" (43).[9] No later historian or administrator has produced such a remarkably systematic and synthesizing study of the first hundred years of British forestry in India.[10]

This taxonomy of forests, with its omniscient divisions and subdivisions; its families, genera, and species; and its attention to geography and topography, is an interlocking whole. It presents a set of nested, hierarchical relationships among constituent units. New forms of colonial knowledge made it possible to build this hierarchical classification from the ground up, even as it was presented in a seamless fashion, beginning with the topmost layers of the hierarchy. Different species and genera of vegetation come together with specific frequencies, and their distribution produces specific vegetal regions and natural zones.[11] Ultimately, this taxonomy facilitates, and is a part of, the application of statistics and numbers that was to constitute India's forests within a new representational regime.

Specific operations of control brought into being the forests that Stebbing describes and classifies. The regulatory elements that went into the making of forests depended on a prior imagination of forests as territorial, exhaustible resources that could be assessed in their entirety in the pursuit of three goals: improvement, conservation, and revenue. Statistical techniques and numerical representations facilitated this way of imagining forests as foresters began to name, count, and measure so as to assess existing vegetation. Strategies to improve the forest estate included new ways to plant, sort, order, and produce vegetation through rational methods of regeneration. To govern interactions between forests and those who relied on forests, they implemented a series of protective regulations in the name of conservation. And they began to harvest, standardize, transport, value, and market timber in a far more systematic fashion so as to increase revenues. The three goals of improvement, conservation, and revenue thus required the invention of a whole series of official procedures and then their incorporation into governmental practice. Though used piecemeal in different parts of India quite early in the nineteenth century, their compilation as part of a systematic whole occurred first in Burma under the leadership of Dietrich Brandis. Statistics and numbers facilitated the later spread of more standardized practices in British India in part because they allowed precise measurement and comparison of the effects of specific procedures.

Assessing Burma's Forests

Trained in Germany, and knighted for his work on Indian forests, Brandis was appointed inspector general of India's forests in 1863.[12] His work in Burma between 1856 and 1862 preceded his appointment as inspector general and critically influenced the development of Indian forestry operations over the next several decades. In turn, the procedures he adopted and synthesized were part of at least three decades of prior efforts by administrators such as Wallich, Tremenheere, Guthrie, and McClelland. On arriving in Pegu, Brandis began with three objectives

([1897] 1994: 108): to protect and improve teak and arrange harvests so as to stay within regenerative constraints and ensure a permanent and sustainable yield; to control human interactions and make the inhabitants of forests his allies; and to produce annual surplus revenue as soon as possible.[13] These objectives, roughly related to the goals of improvement, conservation, and revenue, were translated into a number of concrete procedures. But an assessment of forest wealth was necessary before any of these procedures could be implemented.

To assess the condition of the forest, Brandis adopted the "linear valuation survey" method. In this survey, Brandis and his assistants walked along predetermined transects: a road, a ridge, a stream, or an imaginary line laid across the area of interest. They surveyed the vegetation, focusing on teak since it was the only remunerative timber species at the time.[14] As they walked, they identified the trees, classified them according to their girth, counted them by making notches on pieces of bamboo representing different size classes, and on the basis of this count calculated the amount of timber in each tree class according to established formulas.[15]

Brandis divided the counted trees into four categories: (1) trees of 6 feet and above in girth; (2) trees whose girth was between 4 feet 6 inches and 6 feet; (3) trees between 1 foot 6 inches and 4 feet 6 inches; and (4) trees less than 1 foot 6 inches and seedlings (Stebbing 1922–26). The surveys led him to conclude that the number of trees in the first three categories was nearly equal in forests that had not recently been worked. As a principle of extraction, he proposed that only trees belonging to the first category should be felled. Further, only as many trees should be felled as would be replaced during a year by the growing stock of second-class trees. To ascertain how many trees from the second category would attain the dimensions of the first, he needed to establish a relationship between age and girth. Using information from Bombay, Java, and Madras, he decided that a twenty-fourth of the first category of trees could be cut each year without endangering a sustained output. Using this survey method, Brandis also established the total number of first-class teak trees in all the forests under his jurisdiction. With better information from surveys in Burma, he amended his initial figures, but the principle for harvesting remained the same: cut no more than the annual increment.

This principle of harvesting only the annual increment can be seen as an adaptation of the striking description of a "model forest" presented by McGregor in his *Organization and Valuation of Forests*.[16]

> Imagine a uniformly productive tract, divided into any number (n) compartments of equal value; the first stocked with trees one year old, the second with trees two years old, and so on in an ascending series up to the nth compartment. . . . And let the revolution or age at which the trees of any compartment are to be cut, be n years. The land will then be parceled out into a number of compartments equal to the number of years in the revolution, and each one will be stocked with trees one year older than those of a compartment immediately proceeding it in age, so that there will be a complete series of groups of all ages from one to n years old. If now, all trees n years old, that is those in the nth compartment, be cut and the land immediately stocked with young growth, it is evident that, at the end of twelve months, the group of trees next in order of age, or n minus one year at the time of first cutting, will have advanced to maturity, while the plants in the first coupe will have taken the place of the youngest crop in the series, and the plants of all intermediate compartments have advanced one year in age. . . . The yearly produce thus obtained is, in fact, the annual growth, or interest, of the material standing on n compartments, and is called the sustained yield, and a forest so organized is called a *model* or ideal forest, because it represents a state of things which is theoretically perfect. (Cited in Kirkwood 1893: 39)

No actual forest resembled this model.[17] Tremendous variations marked different patches of land classified as forest. Differences existed in trees on that land, growth of different trees, annual growth rates of different species, composition of species, interactions among different species, influence of insects and other animals, climate, and soils. It is precisely because of these differences that statistics and numbers became important as mechanisms of comparability. If forests could be turned into their models, a single number representing the exact sustained yield would have been sufficient to manage. But differences required a host of measurements and calculations to facilitate comparisons and decisions. Despite the existence of multifarious influences, foresters could ascertain the number of trees, the volume of biomass, and the value of timber in a given area by using statistical averages and standard deviations that depicted the relationship among area, volume, annual increments, and commercial value within a tolerable degree of fuzziness. Planning and the improvement of plans, and therefore the systematic exploitation of useful trees, depended on numbers and improvements in statistical calculations.

Many of the techniques for acquiring relevant information about an area of vegetation and the relationship between numbers representing the forest had been developed and were being used and taught in Europe.[18] Germany and France were especially vigilant in protecting their forests to facilitate commercial harvests of timber (Rajan 1998a). Although Colbert's Forestry Ordinance of 1669 is well known as one of the first centralized efforts to control forest harvests, its implementation was patchy in the extreme, and French forestry began to use statistical methods systematically only in the nineteenth century.[19] And, indeed, management of vegetation using statistical techniques and mathematical relationships did not begin in India until it was imported and adapted from Europe, especially Germany. Our understandings of forests are based on the spread and normalization of these statistical and numerical techniques, and their underlying conception of forests as model entities to be managed more efficiently for human consumption.

But there were significant impediments to importing, appropriating, and adapting specifically European techniques and knowledge base on the Indian context. For one thing, the number of species of trees in any tropical forest was immensely greater than that in European forests.[20] For a method of management that depended on extreme simplification of the landscape, this was a major hindrance. Second, even for the more important species in India forests, there was little or no available information about age, growth rates, and valuation. And, third, although British India had been fortunate all through the eighteenth and early nineteenth centuries in gaining from the knowledge of many noted natural historians, there were few trained foresters who could realize the new vision of forests.

Within the first fifty years of the establishment of various forest departments, however, British administrators and conservators proved themselves equal to the task of addressing many of these difficulties in substantial measure. They focused on specific tree species for their commercial or strategic value and used working plans to affect the composition of plant communities over large tracts of land. They drew on available information regarding specific species and refined it over time with reams of statistical information on relationships among age, variables such as diameter and girth, and tensile strength and moisture content. Reports in the *Indian Forester* and information from various forest manuals began to contribute useful data to assist foresters' management goals by the mid-1890s.[21] Training of Indian foresters began to take place in an Indian school in Dehradun as early as 1878.

Improvement: Consolidating and Reducing Diversity

The acquisition of knowledge about trees, their occurrence, and their distribution was a critical first step in making land and vegetation ready for environmental government. This knowledge facilitated the design of management interventions to protect forests and exclude external influences. It was also crucial in reducing diversity. In Burmese forests, where species diversity was too high and growth rates of hardwoods too low to permit clear felling and fresh planting on a sustainable basis, foresters adopted other measures to enhance efficiency. They encouraged the growth of teak by protecting it against natural obstacles such as creepers, parasites, the shade of other trees, and fire. To give teak more room to breathe and grow, Brandis initiated research into the qualities of other trees in Burmese forests so that they could be marketed. Even if their sale only recouped the costs of harvesting and marketing, their removal freed more room for teak.

To improve the prospects for teak growth, Brandis initiated several other measures as well. These included experimental plantations, the creation of nurseries, and scattering teak seeds in the forest. Large areas with bamboo were cleared by fire, and seeds were scattered over the land to promote the cultivation of teak forests. To counter the high costs resulting from the scattered distribution of teak trees in Burmese forests— few forests contained more than one teak tree out of three hundred and the half a million marketable trees were scattered over 7,000 square miles—consolidation of teak patches was essential.[22] His procedure was to select areas that were suited to the growth of teak and through protection, sowing, and planting increase its density. Higher concentrations of teak justified the construction of roads, timber slides, sheds, and offices.

Conservation: Protecting the Forest and Regulating Human Actions

For foresters, a recurrent problem in efforts to improve vegetation for commercial gain was external human influences—especially by those living close to the forest. Brandis introduced a set of twenty-two rules in Burma, spelling out how to protect forests from humans. The rules established state ownership over forests in the name of conservation and portrayed the forest department as the appropriate agency to implement the rules. Fourteen of these rules regulated behavior with respect to teak. Girdling and felling, cutting or breaking off branches, injuries to seedlings and smaller trees in the process of removing felled timber, fires and clearing of vegetation for *toungya* (shifting cultivation), construction within designated boundaries of forests, methods of obtaining timber for private subsistence use, and procedures for removal of stumps and branches—all of these activities were regulated most strictly where teak was concerned (Stebbing 1922–26: 373–76). For Brandis and his department, forests in Burma existed and were to be regulated to the extent that they contained teak: "Teak does not form pure forests, but is always mixed with a large number of other trees which are either valueless or of little value" (Brandis 1881b: 1).[23] These other valueless trees were slowly to be eliminated.

Part of the effort to control human influences in the forest was the reshaping of the practice of toungya. Rural dwellers residing near dense vegetation felled and fired it prior to cultivating the land for a few years. Brandis's objective was to tame this practice to the ends of forestry. As early as 1856, he encouraged toungya cutters to sow teak seeds and seedlings in regular rows together with the rice they planted. Over the next two decades, this practice grew into a regular system, with many of the toungya areas being stocked with teak at costs far lower than those incurred in regular plantations of teak. That Brandis realized the value of persuading local populations to become accomplices in governing forests is reflected in a remark: "This, if the people can be brought to do it, is likely to become the most efficient mode of planting teak in this country" (quoted in Stebbing 1922–26: 378).

As early as 1876, Brandis was using statistical arguments to defend the demarcation of areas from which humans were to be excluded. Using figures on the yield of teak from forests in Burma and the import and export of timber from Rangoon and Moulmein, he advocated the creation of forest reserves in which toungya cultivation could not occur. Initially he suggested that 1,200 square miles of reserves would be sufficient (1876: 2, 10). This limited ambition did not survive the ero-

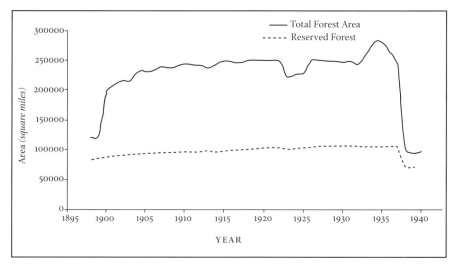

FIGURE 2: Area controlled by the forest department, 1898–1940.

Source: Annual Return of Statistics Related to Forest Administration in British India, 1880–1940.

sive flow of large revenues. The forest department controlled more than two-thirds of Burma's land (29,000 square miles) by 1928, nearly twenty-five times the initially estimated requirement (GOI 1930: 17). With important regional variations, the story was similar for India. Figure 2 provides information on annual change in the area that the forest department controlled, and the extent of reserved forests between 1898 and 1940.[24]

Revenue: Working the Forest

Brandis's preferred method for working the forest was the same as the one adopted in the Anaimalai teak forests in South India earlier in the decade. There Lieutenant Michael, appointed superintendent of forests in 1854, directly employed axmen and sawyers to shape felled teak logs into planks, partially season them in the forest itself, and transport them along a river after they had been slid down a slipway. But for the most part forests in India were worked under far less restrictive conditions. In Bengal, where Dr. T. Anderson was appointed the first conservator of forests in 1864, most forests had been leased to contractors, who felled timber under few specified restrictions. Efforts at departmental working of forests generated difficulties well into the late nineteenth century (Sivaramakrishnan 1999: 156–58).

When Brandis arrived in Burma, he also faced a choice among different methods for working the forests and raising revenues. The existing system was based on auctioning trees to the highest bidder. The winning bidder was responsible for felling and marketing the timber. The system meant less work for the forest department and supposedly encouraged private enterprise. But contractors also exercised significant discretion in felling trees, which often led to clear-felling that conformed neither to the objective of forest conservation nor to the appropriation of significant revenues. And it proved difficult to create a structure of enforceable rules that would make winning bidders harvest only mature trees, plant seedlings, protect trees that were not supposed to be harvested, or construct reliable roads and other infrastructure to facilitate future harvests. An alternative for Brandis was to have the forest department work the forests directly, transport the felled timber to depots, and sell the seasoned wood to the highest bidder. Under this system, the forest department harvested timber either directly or with the assistance of contractors, had it transported to its depots, and then auctioned the seasoned timber to the highest bidders (Stebbing 1922–26: 370–72). This choice about the appropriate method again depended on the department's ability to examine the benefits of different alternatives using numbers and statistics.

Middlemen were crucial in timber harvests, however. In figure 3, note how middlemen purchasers played an especially important role during World War I and the Great Depression.

The most distinguishing and lasting aspect of Brandis's operations in Burma were perhaps his efforts to create a body of statistical knowledge about forests that would be useful for their commercial exploitation and would yield efficient results.[25] The concern for statistical information insinuated itself into the reports of forest departments in the early twentieth century in a manner far different from earlier textual descriptions of Indian forests by natural historians, medical surgeons, and administrators.

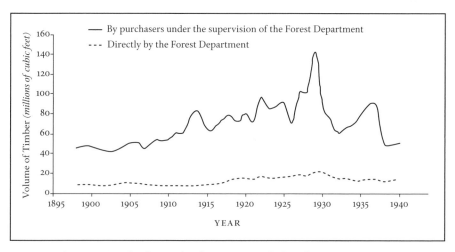

FIGURE 3: Agencies involved in timber harvesting in Indian forests, 1898–1940.

Source: Annual Return of Statistics Related to Forest Administration in British India, 1880–1940.

Forests of Statistics

If one looks for quantitative methods or massive tables of information in Indian natural historical writings in the eighteenth century, one will look in vain.[26] Most of the writings on Indian flora were descriptive. Some were highly detailed classifications in the Linnaean mold.[27] Others were accounts of the country, with special emphasis on its vegetation. Several general publications on Indian flora had appeared early in the process of colonization. Grove tells the fascinating story of the foundational role of indigenous taxonomies and knowledge of plants in the seventeenth century for the later creation of "European texts on south Asian botany" (1998a: 192). John Forbes Royle's *Illustrations of Himalayan Botany*, published between 1833 and 1839, described 207 families of plants and was later described as the first and only attempt to connect features of plants and their distribution in northern India with the elevation and climate they inhabit (Hooker and Thompson, cited in Sangwan 1998: 219).[28] None of these publications relied on numbers or statistics in any significant way.

The more descriptive studies of forests and the country were often produced as exploratory reports, travelogues, notes by traders, official descriptions, letters, and dispatches. Many accounts contained some passing information about forests as part of a larger concern with the resources in an area. But even when they contained quantitative information about other features of the social landscape, such as villages, cultivated land, or major crops, little similar information about forests was available. Most reports on forests during the early eighteenth century gave only a general sense of the important features of the vegetation in an entire region and in some cases saw forests only as obstacles to the extension of agriculture.[29]

By the early to middle nineteenth century, many of the reports written by natural historians, medical surgeons, and administrators described high levels of wasteful tree felling by private contractors and timber traders and the effect of such felling on changes in local climate and rainfall patterns. These were again highly descriptive accounts with little quantitative information. Instead of numbers, we have texts filled with details about trees, their characteristics, and the uses of specific varieties. Equally prolific are speculations about the effects of deforestation and the possible uses that forests might have. But rather than precise quantitative fixes or estimates of ranges we learn that forests are dense or sparse, their extent is vast or limited, and their uses are many or unknown. Numbers did not go together so well with forests or their characteristics (Lloyd and Gerard 1840; Vigne 1842). Even when these writers described specific interactions among plant communities, they did not use numbers to characterize the relationships (Forsyth 1872). And when some of the authors provided measurements of trees during their travels, like Nathaniel Wallich did in Burma, it was less an attempt to assay the forest itself than a means of portraying outstanding specimens of valuable species (Stebbing 1922–26: 139).

These ways of thinking about vegetation began to change in the latter half of the nineteenth century. The relative trickle of writings on natural history turned into a flood of scientific facts, figures, and statistics.[30] The crucial difference in these publications, compared to descriptions of an earlier age, was their greater emphasis on precise quantitative information to illuminate forest utilization and protection, forest surveys, preparation of working plans, silviculture, and forest policies and regulations.

This later period also saw the birth of a large number of periodicals devoted to forestry. William Schlich founded *The Indian Forester* in 1875. The different provincial forestry departments prepared regular working plans and published annual reports that condensed the chief features of their operations in the form of statistical tables. With the creation of a Forest Research Institute in 1906, periodicals such as *Forest Bulletins, Forest Pamphlets, Forest Leaflets, Forest Records, Forest Memoirs,* and *Forest Manuals* became required reading for those interested in forestry. Pocket-sized books that foresters could carry conveniently provided yield, volume, outturn tables by species, and masses of statistics to help convert a given log of wood into its content in cubic feet and a day's wage into monthly or annual wages (Howard 1937). Provincial forestry services also published their own series of specialized pamphlets and bulletins on issues of regional interest.

The first *Review of the Forest Administration* in India after the passage of the Forests Act of 1878 contained quantitative information, mainly on the area of land under the control of the forest department, and the revenues and expenditures of the provincial forest departments (Brandis 1879). There were other tables scattered throughout the report, but the nature of the information was more summary than thorough. It hinted at the quantification that was to come, especially in the everyday operations of the forest department, with the embracing of working plans for increasing areas of land, survey operations to demarcate and bound departmental forests, and adoption of mensuration techniques for more and more timber species.

Quantification was necessary for the desired uniformity in provincial and divisional operations. But it faced significant difficulties. Brandis summed up the tension well: "Forest management must always be essentially local; its operations are governed not only by the peculiar climate and character of Forest growth of each district and province, but equally so by demands of trade, the land tenures, and the customs regarding the use of the waste. . . . Yet on the other hand Forest administration in all provinces and district is, or ought to be, governed by the same general principles, and hence a comprehensive review . . . cannot fail to be useful and instructive" (1879: 1).

Despite difficulties and periodic modifications in reporting formats, the annual reviews of forest administration provide a good sense of how "a comprehensive review" depended on statistics. The report for 1903–4, for example, is sixty-four pages long of which the last thirty are devoted to twenty-four tabular appendixes of numbers describing forests in each province. In the textual part of the report, there are additional fourteen tables filled with numbers. Together, these thirty-eight tables yield a bird's eye view of land classified as reserved, protected, or unclassed forest; changes in the area of these classifications; the progress and nature of survey operations and boundary demarcation; the adoption of working plans; breaches of forest rules; types of offenses committed in relation to forests; protection from forest fires and grazing; area devoted to plantation activities; output of different forest products; agencies involved in forest exploitation; the value of forest products harvested; and detailed budgetary statements on revenue, expenditures, and surpluses. The tables, appendixes, and maps bear witness to the insufficient, unceasing production of numbers, insufficient because more numbers could only improve representations, never perfect them, and unceasing because the lure of better, more accurate representation provoked the production of even more numbers.

By almost any measure, the numerical representations were the result of a remarkable feat of synthesis carried out year after year. They provide a sense of the staggering results that the forest department achieved. Over the five years between 1899 to 1904, the area under the direct or indirect control of the forest department grew from 122,000 to 232,000 square miles.[31] Looked at another way, in these five years the forest department almost doubled the amount of land it controlled: 24 percent of India's territory, up from about 13 percent. Other figures are similarly impressive. The surplus of 10 million rupees in 1904 was just about half of the total revenue of 22 million for the forest administration. The 10 million rupees represented a nearly fivefold increase during the preceding thirty years. In 1874, the department had generated a surplus of just 2.2 million (Eardly-Wilmot 1906: 2, 26). Figure 4 shows the increasing financial scale of the operations of the forest department: it enjoyed a surplus of revenues over expenditures in every single year during these six decades.

Despite the high level of exploitation of India's vegetation by the department, there were strict checks on subsistence use by local residents. The department's reports present information on breaches of forest rules (through fires, unauthorized felling, appropriation of non-timber products, and grazing), whether the case was resolved, and whether the identity of the offender was known. Forest officials claimed

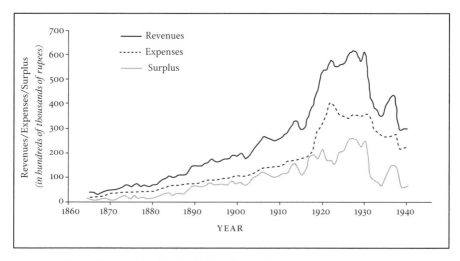

FIGURE 4: Financial statistics for the Indian forest department, 1865–1940.

Source: Annual Return of Statistics Related to Forest Administration in British India, 1880–1940.

to have detected nearly sixty thousand offenses throughout India. More than 80 percent concerned grazing and removal of forest products—activities rendered illegal by the new regulations. But these offenses pale in comparison with extraction by the forest department.

Even if breaches of rules were a cause for concern for the department, officials were baffled about what to do. The 1906 review asserts: "It is impossible to give reasons for these statistical variations. The case of each Province or even District presents different conditions which Local Governments treat, after due investigation, with punitive or alleviating measures suited to circumstances" (Eardly-Wilmot 1907: 9).

The foresters' bafflement when forced to explain variations in social pressures on forests was at odds with their ambition and accomplishments where more biological aspects of control were concerned. Observations and experiments provided the basic information necessary to manage vegetation. Surveys, demarcations, and settlements of land prepared the ground on which this information could be used. Learning lessons from experiments and surveys was typically possible because numbers could be used to represent results, which were then utilized in the implementation of plantations and working plans.

By the 1890s, experiments that examined the effects of various methods of propagation, growth, and management of indigenous and exotic species were under way in most parts of India. Results were disseminated through a variety of vehicles, including the *Indian Forester*, but especially through special publications such as the *Forest Bulletins* and *Forest Records*. The bulk of the experiments focused on timber, with a significant emphasis on ascertaining the germination rates, rates of growth, and area to volume conversions for various timber species. Each year, the forest department also undertook survey, demarcation, and settlement operations for the land it controlled. By 1905, the department had surveyed nearly half of its possessions. Surveys were followed by demarcation and settlement of the areas as reserved or protected forest. Careful accounting revealed the progress (or its lack) in each of these domains by providing precise, if not always accurate, numbers.[32]

Statistics and Plantations

The influence of numbers and statistics in imagining and creating a new type of forest was perhaps most visible in the forest department's plantations. The overall area of the department's plantations was always a small proportion of total area under its control, but plantations are still important for two reasons.[33] First, actions in plantations are the most comprehensive official interventions, and far more careful records were kept than for areas that were already vegetated. Second, when successful, plantation forests produced far higher levels of revenue and reliable information than forests managed primarily through clearing, thinning, and felling, and without much recourse to sowing and plantation work. Consider as an example the Nilumbur plantations in Madras, one of the earliest efforts to create a plantation forest. It was calculated that the value of teak planted there between 1842 and 1850 on approximately 700 acres would be more than 1.1 million rupees in 1878 (Beddome 1878). Compare this figure with the surplus of 2.2 million rupees for the entire Indian forest department in 1880 (Ribbentrop 1896: 70)! Actual revenues came nowhere near these optimistic estimates.[34]

The district collector of Malabar, Mr. Connolly, initiated the Nilumbur teak plantations in 1842.[35] To guide forestry practice, Connolly specified seven types of rules which resembled some of the prescriptions for protection that Brandis was to create in Burma a decade later. These rules addressed issues of planting, inventory, felling, contractual arrangements, monitoring, enforcement, and personnel.[36] Chatter Menon, a "native," served as the subconservator of the plantations until his death in 1862. His methods for germinating teak proved so successful that they continued to be used for another half century.

One of the most comprehensive records of the status of the Nilumbur plantations is available in an 1878 report prepared by the conservator of forests at Ootacamund when the government of India sent several teak specimens from the plantations to an exhibition in Paris (Beddome 1878). This report provides information on the growth of teak in the entire plantation. With information on every tree above a certain size, the report is invaluable as an indication of the care lavished on the plantations by the department. The complete exclusion of fire, cattle, and creepers had led to "wonderfully straight [trees] with only one stem. . . . The immense length of the boll in every tree [gives] the plantations an enormous advantage . . . since the yield per acre will be out of all proportion to that in any natural forest in India" (26).

For the argument in this chapter, the most interesting part of the report is its statistical appendixes. The forest department considered a density of sixty trees to the acre the most desirable. Beddome lists precise measurements of trees for several half-acre plots from different years of planting to demonstrate the remarkable profits the plantations were likely to yield. We learn that teak begins to amass girth (the source of greatest value) only after the first twenty-five years of its life, then increasing by 200 to 400 cubic feet per acre in a single year. Beddome's measurements reveal the importance of numbers, averages, variations from the average, and mathematical equations in calculating the volume of wood on a given plot of land. Once diversity in a forest was minimized—the overriding objective in plantations—and the volume of wood yield over time was known, the calculation of revenues and profits, decisions about future plantations, and the timing of harvests and transportation to market were simply a function of prevailing prices.

Statistics and Working Plans

New manuals of silviculture and forestry facilitated the use of numbers and statistics in departmental work. With their yield tables and descriptions of standard procedures, they helped stabilize expectations about "normal" conditions and also provided a basis for assessing the extent of deviations. They were extremely useful in creating new forests but invaluable in shaping the contours of existing vegetation. The need for complex calculations using numbers and mathematics was obviously greater for vegetation that had not been planted at a known time and whose growth rates therefore could be estimated only by positing rough relationships between girth and age. Assessments of market value of acquired landscapes, where trees of many different species were mixed with a vast and undetermined assortment of other vegetation, needed an immense labor of identification, classification, and enumeration. To this end, the forest department prepared detailed working plans to consolidate its holdings and manage the vegetation on the lands it came to control.

The rate at which different parts of India came under working plan management varied greatly. In 1908, 90 percent of the forest land in the United Provinces and all of it in the Northwest Provinces was being managed through working plans. In contrast, working plans had been prepared and sanctioned for only 4 percent of the forest land in Burma, eastern Bengal, and Assam (Beadon-Bryant 1910: 33). Although all working plans shared a core of common features and similar organization of materials, they did not all contain precisely the same information, especially in their numerical appendixes.[37] Some of them focused more on describing the existing state of the vegetation or on the supplementary regulations necessary to control and reshape the existing vegetation. Others focused on the means by which such forest making could be accomplished or on the anticipated outcomes. Further, working plans underwent modification depending on the extent to which anticipated outcomes materialized.

Variations in the contents of the plans depended on the relative importance of the management objectives for a particular patch of land: firewood production; prevention of soil erosion; rationalization of existing diversity; and, most importantly, higher timber yields. Variations also resulted from the differing conditions across space and the differing histories of government over various patches of vegetation. Due to differences already in existence when the forest department became part of the history of a given landscape, the timing and nature of its entry itself produced variations. It was precisely because existing variations had to be tamed that working plans differed in their contents and prescriptions. But the common objectives of improvement, conservation, and revenue also lent them a common organizational core.

Each plan was based on several visits to the field by foresters. Plans were typically divided into two parts. The first described the situation as it was. This part contained information about climate, soil, topography, geology, size and boundaries of the area, condition of the vegetation, species composition, distribution of the existing "crop of trees,"[38] systems of management and property, levels of products, quantities and value of harvests in the past, local customs and agrarian practices (where relevant), and how the area was connected to markets and means of transportation. Where possible, maps revealed the spatial relationships among the different parts of the area.

Quantitative information was important in ascertaining the situation on the ground in a number of ways. Foresters used numerical tables to classify a given area and its vegetation on the basis of a whole range of factors. Take as examples just three elements: climate, soil, and vegetation. Information on different aspects of each of these elements was available in numerical tables that facilitated estimates of the potential performance of a given patch of forest. For climate, foresters might consider rainfall, temperature, light, and moisture; for soils, they might examine mineral composition, organic matter, water retentivity, consistency, permeability, and depth; and for vegetation they might focus on morphology, space requirements, age, height and growth relationships, volume, life expectancy, means of reproduction, and diversity. An analysis of the forest in terms of numerical tables made them accessible to intervention and management as relationships among these numbers became clearer with successive experiences over time. The tables revealed variations from the norm and allowed informal comparisons across plans at a glance. Numbers were crucial to the success of each plan because they allowed foresters to gauge the type of management intervention necessary in a given location and to refine the scope, intensity, and location of interventions. In the absence of numerical representations, and knowledge about how different kinds of numbers related to each other, the forests would have been managed on a far cruder scale.

The second part of the plan concerned a grid of legibility and control, which would make the existing features of the landscape match the goals that were the basis of any working plan. Department officials divided the selected area into working circles, each with an area ranging from a few square miles and more than a hundred. The type of vegetation and the extent to which it would require similar treatment before timber could be extracted usually determined the creation of a working circle. These circles were further divided into compartments based on natural landmarks within the forest so that divisions would be well bounded and easy to harvest each year. Vegetation within each compartment and circle was then classified by size, trees in each class were roughly enumerated, and the total was summarily valued. Forestry officials counted, measured, and valued only the most dominant and valuable commercial species in the compartments.

Once foresters determined these numerical facts, methods for treating vegetation and felling timber followed more or less as a matter of course. Knowledge about rates of growth and size-class distribution of timber across the compartments influenced the timing of treatments and fellings. The plans proposed rotations anywhere between ten and forty years for working a compartment. Since trees in a compartment that had never been worked did not conform to any particular size class, it was often necessary to return to the same area to reduce its size-class variations. Precise numbers on past results and historical rates of growth helped refine future proposals.

In addition to prescriptions for treatment and felling, each plan enumerated other activities and restrictions for a working circle. These concerned planting, thinning, clearing, and improvement-felling operations to amplify tree growth. To check external influences, plans stipulated controls on grazing, fodder harvesting, firewood collection, removal of subsistence timber, and fire. Such additional restrictions set forest department activities and staff on a collision course with villagers and others who depended on forests. The first forest conservatorship, in the early nineteenth century, was abolished precisely because of the ire it produced among villagers, landowners, and merchants. Later forest operations in Bombay Presidency led to letters of complaint signed by more than eighteen thousand people and the appointment of the Bombay Forest Commission in 1885 (see Bombay Forest Commission 1887–88). Similar results were also visible in the United Provinces in the early twentieth century.

Working plans wove together the strategies of control that foresters needed to reshape vegetation and create forests fitting the objectives of efficient and convenient exploitation. The magnitude and complexity of the tasks involved in creating even a single plan exposes the reliance foresters placed on numbers to learn from the past and propose for the future. Classification and enumeration of all valuable trees, and a schedule of harvests stretching fifty years into the future, would simply have been impossible in the absence of statistical techniques, expectations about averages, and projections about future prices of timber. Before colonialism, they were. Statistical histories of results, quantified information about production levels, and numerical assessments of pitfalls made working plans and forests possible.

Of course, no working plan came to pass as conceived initially. They needed constant modifications in light of experiences with communities of vegetation and people. All plans are fated to be revised. But the point is not whether events and outcomes conformed precisely to plans. It is rather that foresters applied numbers to vegetation in ways that made planning possible. It is that the use of statistical techniques led to unprecedented modifications of forests and understandings of forests. It is that the role of numbers and statistics in the creation of Indian forests needs greater attention.

Forests and the Rule of Numbers

> Before the end of the eighteenth century, *man* did not exist. . . . He is quite a recent creature, which the demiurge of knowledge fabricated with its own hands less than two hundred years ago. . . . When natural history becomes biology, when the analysis of wealth becomes economics, when, above all, reflection upon language becomes philology . . . then, in the profound upheaval of such an archaeological mutation, man appears.—Michel Foucault, *The Order of Things*

The previous sections have documented the centrality of numbers and statistics to the strategies of knowledge/power on which the colonial government of Indian forests was founded. The purposive selection of a set of features that could be statistically represented—among them, number, area, value, and percentage tree cover—helped define Indian forests and facilitated interventions to alter their shape. No prediscur-

sive understandings of that set of objects called "Indian forests" survived the history of selective representation and construction launched
in the nineteenth century. Contemporary meanings of forests and their
place in the natural heritage that Indians need in order to conserve and
protect them, manage and sustain them, and pass them on to future
generations have become possible only through widespread and enduring recourse to statistics. The standardization that numbers introduced
in representations of the occurrence, density, health, and diversity of
forests continues to shape understandings of forests today.

Colonial interventions to shape landscapes and bring forests into
being were founded on an underlying conception of forests as valuable
resources that required demarcation, exclusion, and government to
yield optimum levels of production; they institutionalized the science
of forestry through new organizational mechanisms and technical refinements that continue today; and they relied on numbers and statistics to represent and reshape forests.[39] Numbers and statistics achieved
their special force in conjunction with the other two interventions but
simultaneously helped validate them. In this sense, one can truly refer
to a mutual constitution between statistical representations of forests
and what forests came to be and mean.

Representation by numbers transformed beliefs among foresters
about ideal forests and made possible the reworking of existing vegetation in terms of scientific forestry, sustainable yields, and profit maximization. Just as a later technology of government would help produce
new environmental subjects among communities (chapter 6), the centralized government of forests helped shape the views of foresters and
official employees. Working plans, forestry and silviculture manuals,
and administrative reports depicted forests using a host of different
kinds of numbers. Such numerical representations strengthened technologies of government by facilitating four types of operations in forests. The first constituted forests as a domain fit for government. The
Forests Act of 1878 defined forests simply as a legislated category of land,
not in terms of trees, plants, soils, or vegetation cover. A forest was just
what the government said was a forest. Only by appreciating the confidence of foresters that given the right conditions they could create a
forest on bare land does such a legal definition make sense. Surveys,
demarcation, the division of land into particular categories of protection, and treatments based on working plans made up forests. Foresters

identified and enumerated individual trees in a landscape; classified these trees into spatially fixed statistical categories tied to age and size, abstractly aggregated these statistical categories on the basis of yield, valuation, and projected revenues; and converted each category of trees and land into a volume of wood and its likely market price. Numbers and statistics made it possible to constitute forests.

New forms of statistical knowledge also contributed to forest making by helping create a new history for them, a history free of the commonplace and extraordinary problems to which all landscapes and vegetation are typically subject. Instead forests were characterized by performance on indicators of yield and revenue. With the implementation of a working plan, a forest commenced a new life in which the performance of spatial divisions was at stake rather than their uses for local residents, their existence as a community of vegetation, or their aesthetic import.

Numericized knowledge helped in a third way. It gave concrete form to potential obstacles to the conservation and growth of forests. Statistical government of forests aimed at the precise calculation of the effects of such obstacles to steady growth as fires, insects, parasites, creepers, agriculture, grazing, firewood collection, unauthorized extraction of forest products, and careless harvesting of timber—in short, any of a host of events that could limit or threaten those "wonderfully straight [trees] with only one stem" (Beddome 1878). Silvicultural knowledge based on field experiences, and later codified and disseminated in forestry manuals, indicated how, when, at what intervals, and with what results various methods of treatment could be used to overcome such obstacles to production (Schlich 1910: 92–117). Numbers also played a critical role in that they helped pinpoint deviations from the norm more precisely. Efforts to execute working plans generated resistance based on climate, soils, history, and human uses. The desire to achieve goals required accommodation.[40] The interplay of implementation, resistance, and accommodation was unavoidable in a context where the unceasing generation of statistical information was always insufficient owing to the inherent complexity and unpredictability of the factors that affected growth and output. Foresters addressed these factors to achieve desired numericized goals of production from a given area of land; these goals were based on norms established statistically, taking into account the effects of local variations.

Finally, numerical understandings of forests allowed foresters to establish commensurability in their actions and results. They could use numbers to compare the performance of forests and forest departments anywhere in India by using uniform criteria such as rates of growth; relationships between the age, girth, and length of trees; volume of wood or other biomass; revenue streams; and so forth. Even when they recognized the importance of local and regional specificity, the use of statistics to represent the most valued features of forests and timber transformed them into comparable entities.[41] The dimensions of costs, yields, and surpluses were relevant for all land and all trees, and by the end of the nineteenth century summary representations of Indian forests had become routine. Changes in statistical indexes denoted performance over time. Statistical measurements in annual reports, budgetary records, bulletins, research results, and working plans were basic to the self-assessment of provincial forest departments, crucial to the exercise of central guidance, and constitutive of the relationship between knowledge and government. They helped create a new generation of foresters who had faith and confidence in the technologies of government of which they were a part.

Each of the four ways in which numbers helped make forests had a technical aspect. But each is also obviously political in nature. What Ludden says about the role of colonial knowledge in the construction of a traditional village India (1993: 259–63) is just as relevant for the role of colonial knowledge in the construction of forests. The choice of certain characteristics of landscape and vegetation to represent forests numerically was based on prior selection of the objectives of timber production and revenue maximization. The implementation of exclusionary techniques to enhance growth displaced many people from their lands. But numbers gained the forest department the extraordinary triumph of seemingly eliminating politics from the government of forests. After all, who could argue with the assertion that 200 cubic feet of wood per acre per year were better than 100 (or even 199)? Efficiency in all its forms now had a concrete content.

In the ease with which numbers could be used to organize and rank outcomes lay one of the secrets of their success as "apolitical inscriptions." As Rose puts it, when "numbers are used as 'automatic pilots' in decision making, they transform the thing being measured—segregation, hunger, poverty—into its statistical indicator, and displace political disputes into technical disputes about method" (1999: 205). The development of systematic procedures of identification, enumeration, classification, calculation, and valuation through working plans was an attempt to eliminate or at least reduce the influence of the individual in governing forests: idiosyncratic and subjective effects of individual preferences and biases had no place in a systematic and rational government of forests.

Ultimately, depoliticization of environmental government depended precisely on the ability of forest department officials to model the forest effectively as the product of a small set of variables and relationships. By limiting the representation of desirable features of forests to one or two or three numbers, and by yoking the governmental practice of entire generations of officials to the maximization of these numbers related to volume, growth increment, and revenue, other products that forests could provide became devalued. To the extent that firewood, fodder, and household timber were necessary for the livelihoods of many persons living close to forests, they might grudgingly be admitted into the calculation of foresters, though only as a constraint on the main goals of forestry.

Those who objected to the forest department's efforts to control ever increasing areas of land in the name of state interests, the environmental health of the country, or greater revenues also used numbers in their arguments. Numbers were deployed in the rivalries between the revenue and forest departments in Bombay, Kumaon, Madras, and Punjab. Although government officials desiring greater control over land were the first to use statistics, opponents of centralized control also began to question, undermine, and reshape existing policies on the basis of statistical information as it became more widely available. Where government is concerned, statistics displaces verbal descriptions.

Conclusion

In trying to trace the making of forests in nineteenth-century colonial India, this chapter has focused on the role of statistics and numbers in new technologies of government. It has assembled some of the chief instruments through which new views of forests came to be stabilized and won for their proponents control over nearly a quarter of India's territory. Numerical tables, summary figures, and statistical relationships were pervasive in writings on forests by the last quarter of the nineteenth century and made new views of forests possible. Foresters began to see statistical tables about forests as precise, apolitical representations of their estate, and they worked to enhance their performance in ways that would be reflected in their numerical reports.

But the emergence of these forms of representation and the specific ways in which forests came to be made through numbers were not inevitable. A particular way of imagining and governing forests came to prevail because of the location of timber and trees in the colonial political economy, the initial investigations of a large group of natural historian-administrators, and the formation of an organizational frame with a long-term interest in appropriating large areas of land for timber. Statistical tables became the best way to represent forests, techniques of government, and the effects of these techniques because of additional factors, including the temporal sequence of colonial conquest and the emergence of statistics in the early nineteenth century as a means of representing social phenomena. It is in the intersection of such varied processes that the contingent nature of all representations and their consequences is to be found.

Nor is it the case that some underlying material reality of a physical entity that we call forests made it inevitable that its representations would assume the final form they did. Forests have been represented in many ways, some of them radically different from each other (Harrison 1992). Even the stability of current representations is questionable. It is likely correct that perceptions of forests as resources that serve human ends and can be governed more effectively through numerical representations and managerial techniques underlie most contemporary actions in forests regardless of whether human actions are concerned with livelihood, profit, or aesthetic preservation. But the persistence of these perceptions is as much a result of the uses to which they are put and beliefs about their scarcity as it might be the result of some inherent qualities of forests or of their internal structure.

In becoming durable devices of representation, statistics and numbers contributed to the creation of forests through a double erasure. They contributed to the way forests could be imagined by reshaping the policies affecting the form and characteristics of forests in national life, by changing the lens through which people viewed them, and by introducing a new language through which to imagine them.[42] New understandings of forests hinged on the possibility that some prominent aspects of existing vegetation could be selected and signified by means of counting and calculation. Statistics accomplished a shift in the context of the objects being selected for representation (Cohn 1982: 301). The numbers representing selected features of the forest then erased, and came to stand in for, the vegetation from which the numbers were abstracted.[43] In the case of Brandis's work in Burma, the number of teak trees in different size classes initially represented what Burmese forests meant to the forester. The erasure of an ensemble of vegetation and relationships, both human and nonhuman, was necessary for forestry to proceed apace. The most significant characteristics of the forest became the number of specimens of preferred species, their age class and size structure, the rate of growth or annual increment, the volume of timber that could be harvested, and so forth. Scott (1998) provides a stinging criticism of such "state simplifications." Numbers about a forest could come to stand for it, and the relationships among these numbers could become the basis for future proposals about what to do in it. The constituent elements to which the vegetation was reduced could be recombined and restructured in light of desired goals to produce new forests.[44]

This re-creation of ideal forests on paper and in plans, and the consequent implementation of the planned paper forests in preference to existing communities of vegetation, constituted a second erasure involved in the making of forests. Classifications of inanimate objects such as trees and vegetation types set in motion a very different process of social world making in comparison with the classifications of human beings on which statistical counts are based. When humans are classified, the classification and the allocation regimes of which the classification is a part generate a new politics. Awareness of the classification on the part of the classified person itself may be crucial in the embracing or rejection of the new classification and its political effects. Indeed, a large literature on contemporary social conflicts in India has related their origins to processes of classification initiated under colonial rule, thereby testifying to the durability and inventiveness of colonial knowledge (Dirks 1987; Inden 1990; Prakash 1999).

Thus, in the making of forests, the second erasure was related not just to a particular way of self-regarding and self-knowing but also to the object of government itself. It was embodied in all the plantations and working plans to transform landscapes that the British created, implemented, and bequeathed to their Indian successors. Actions based on plans, yoked to an imagined forest, with its regular increments and revenue streams, seldom came to pass as anticipated. They met local obstacles and underwent constant reformulation. They were the product of a representational economy, but they had to accommodate the dynamics of a political economy as well. It is in the intersection of representations and their referents that the story of forests in India makes the greatest sense.

3

Struggles over Kumaon's Forests, 1815–1916

It is time for us to ask history to tell us who we are instead of beating its sides once more in order to extract a final drop of prophetism.
—Jacques Donzelot [1977] 1997

The first two decades of the twentieth century were a transitional phase in the government of forests in Kumaon. This period witnessed the conjuncture of a range of struggles over forests: between different departments of the colonial state, over different conceptions of forests and the uses to which they should be put, and among state officials, local elites, and common rural residents. Forests were not just the stakes in these struggles. They were also the basis of an unprecedented historical compromise. They became locations that state officials and rural residents formally began to govern jointly in their efforts to care for nature. Crucial to governmental changes were calculation and care and the extension of caring calculation to the level of the community. From being the driving concerns of the forest department, calculations of area of land, number of trees, amount of products that could be harvested, and strategies to shape individual behavior and beliefs became the concerns also of a widening circle of Kumaon's residents and communities.

Through much of nineteenth century and before, forests had been impediments to agricultural extension in Kumaon. Their desired fate was exploitation and clearance (Traill 1828). In the early 1820s, Boulderson, the magistrate of Bareilly, said about forests in his area that "everything which had the breath of life instinctively deserted these forests and not so much as a bird could be heard or seen in the frightful solitude" (quoted in Tolia 1994: 22). It was another half century before forests began to need protection. Still further in the future was the birth of the Chipko movement, with its hallowed place in imaginings around resistance to coercive state policies, nature's fragility, and human struggles to protect it.[1]

Forests became increasingly important in Kumaon's regional economy as the nineteenth century yielded to the twentieth. Many forest products acquired new value—timber, resin, and firewood among them. Technological innovations, experiences of scarcity, articulation with larger markets, and the invention of new needs were some of the factors at play. Merchants, timber contractors, different departments of the colonial government, and local populations and their leaders devised new procedures through which to appropriate newfound value. Their interactions created enduring conflicts that were usually resolved only provisionally. However, the first two decades of the twentieth century witnessed the emergence of a new equilibrium in Kumaon around resource allocation. Under this equilibrium, the forest estate that the colonial forest department had created over the past five decades was partitioned. The claims of communities and state actors were settled, with each receiving one part of the divided estate. At the same time, new mechanisms of rule and government produced quite different relationships among the major actors interested in forests and their products.

Two ironies haunted the apparatus through which states and communities came to care for nature. The more threatened nature was perceived to be, and the more humans intervened to care for it, the more difficult it would be for nature to survive in a realm imagined as separate from humans. Careful calculation on the part of the state and the community transformed nature into the environment.[2] The romantic appreciation of nature as a place to escape human activity was doomed progressively as humans themselves became the agents for saving it. The community and the state entered a new partnership to save and protect the environment by assessing and evaluating forests and refining mechanisms of governance.

If nature could not survive calculated care, the separation between community and state did not survive their emerging partnership to govern forests through community-based conservation. Communities, by definition, must be central to community-based conservation. But the communities that came into being in Kumaon in the early twentieth century were shaped very much by state efforts. The second irony, in part semantic, involves the processes that eroded the separation between the state and the community. This chapter traces the historical processes through which Kumaon's forested environments were made. Locating Kumaon's forests in history, the chapter explores the technologies of government that were central to their environmentalization. It thus points to the changes in human/nature relationships that led to the founding of village-level forest councils in Kumaon.

Forest Regulations in Kumaon

Although forests may have been crucial to the rural local economy of Kumaon prior to the nineteenth century, they played a relatively limited role in the formal monetized economy of the region.[3] The Chand dynasty ruled Kumaon for nearly a thousand years before the British arrived in 1815. It derived most of its revenues by taxing agriculture, trade, mining, and labor. The Gorkhas of Nepal ruled Kumaon for a short period (1791–1815) between the Chands and the British (Pande 1993; Regmi 1999). During this period as well, timber and other forest products formed just about 2 percent of all realized revenues (Atkinson: vol. 3, [1882] 1973: 456, 462).[4] Even these limited revenues from timber and nontimber products were derived from taxes on exports, not as income from a systematically governed resource.[5] During this early period, "the [forest] products consumed within the hills by the people themselves were too inconsiderable to be taken into account" (vol. 1, 845).

The Extractive Phase, 1815–68

The British East India Company conquered Kumaon and Garhwal in 1815. The region extending from the territory east of the Alaknanda River to the border with Nepal came to constitute British Garhwal and Kumaon (Farooqui 1997: 5–6). Kumaon's first commissioner, G. W. Traill, reorganized the local civil, police, and revenue administrations with a relatively free hand. During Gorkha rule, Kumaon's regional economy had declined greatly owing to high taxes on trade and agriculture.[6] Traill simplified civil administration, eliminated many official positions, carried out a new land revenue settlement, and reduced levies that Gorkha rulers had tried to impose (Tolia 1994: 18–32).[7] The importance of Kumaon at this time lay more in its strategic value than in the revenues it could afford the British. Such considerations made Traill's task easier. But even with a lenient hand he doubled Kumaon's land revenue between his appointment in 1818 and his retirement in 1835.

Traill's main contribution to the government of forests was to assert state control over all areas from which timber was extracted. At this time, government income from forests was based on a contractual system. Contracts could assume two forms: areas of land could be farmed out to landowners or tax collectors after the payment of specified dues or they could be leased to private timber contractors, who paid the government a fee on the timber they harvested. As Atkinson observes: "No attempt was made to enforce any system of conservancy" ([1882] 1973: vol. 1, 849). The reason was prevalent perceptions about the vast extent of Kumaon's forests. Why manage a plethora when there are no intimations of scarcity?

Batten, who succeeded Traill as Kumaon's commissioner (1848–56), discussed the supply of wood for mining operations in Kumaon (Atkinson [1882] 1973: vol. 1, 856). He concluded that local forests could provide "sufficient charcoal for the largest English furnaces for a hundred years . . . renewing themselves without limits" (quoted in Shrivastava 1996: 120–21). Looking back on this halcyon period of abundance, one forest service officer observed: "At first there was an idea that there was unlimited forest wealth in India and for years nothing was done to protect forests in any way" (Bailey 1924: 189).[8]

The perceived vast extent of vegetation wealth in Kumaon itself precluded systematic governance. But by the second half of the nineteenth century a combination of factors conspired to increase the perceived value of timber and trees. Demand for timber as wooden sleeper ties escalated rapidly with growth of the network of Indian railways after the 1857 revolt (Guha [1989] 2000). With the rise in the value of timber, speculators became actively involved in its extraction. The government initially leased areas to contractors, "who had uncontrolled liberty to cut where and how they pleased with the result that large numbers of trees were felled and for want of transport were left to rot in the forests" (Walton 1911: 11).[9] Within a few years, Kumaon forests suffered a paroxysm of felling so intense that their first official surveyor in 1869 was forced to conclude that the forests "have been worked to desolation, but perhaps even this does not given an adequate idea of the waste that has occurred and the mischief that has been committed. Thousands of trees were felled which were never removed, nor was their removal possible" (Pearson 1869, cited in Shrivastava 1996: 136).[10]

Tree species such as sal (*Shorea robusta*) were especially valuable for the durable sleeper ties they provided. Sal had always been important commercially. Traill had classified it as a reserved species as early as 1826.[11] But in the latter half of the nineteenth century species such as sissoo (*Dalbergia sissoo*), tun (*Toona ciliata*), and khair (*Acacia catechu*) also became valuable for construction, industrial consumption, and firewood. The use of creosoted pine, a technology developed in England, made pine sleepers more durable (lasting 16 to 18.5 years) than even sal and teak (which lasted 13 and 14 years). This made pine, which grew abundantly in Kumaon, a highly valued commodity in preference to oak and other broadleaved species (Molesworth 1880). Pine also became important because it yielded resin, and in the early twentieth century forest department researchers successfully extracted turpentine and rosin from pine resin (Smythies 1914). Profits from the sale of resin began to account for nearly the total net revenue of the forest department in the United Provinces, especially during the war years.

Henry Ramsay, who succeeded Batten as the commissioner of Kumaon, abolished the contract system for exploiting timber in 1858. He also launched the first forest conservancy in Kumaon. In the initial stage, the organization of forestry in Kumaon differed significantly from that in other parts of the country, such as Burma or the Central Provinces, in that an official from the revenue department was also in charge of forest governance (Fisher 1885: 586).[12] When this arrangement changed and a professional forester replaced Ramsay, the seeds were sown for a long-term rivalry between the forest and revenue departments.[13]

Ramsay implemented a series of measures that slowly increased the area of forests under state control in Kumaon and the type of regulations that the state exercised. He began clearer demarcation of forest boundaries, introduced rotational working of forests, and enforced the practice of hammer marking trees to be felled (Weber 1866: 3). As Smythies observes: "The farming of leases and the indiscriminate felling of trees was stopped. . . . The most valuable sal forests were formally demarcated as reserved forests under the Forest Act of 1865."[14] These steps led to a steep rise in revenues from forests, with an average surplus of nearly 70,000 rupees during 1859–68 (Walton 1911: 12). Forests were beginning to assume a critical role in the regional economy. State revenues from taxing timber and nontimber products had seldom been more than 10,000 rupees a year in the first half of the nineteenth century. In contrast, land revenue was yielding 125,000 rupees by the middle of the century (Atkinson [1882] 1973: vol. 3, 478, 486). The picture changed drastically as state income from forests came to rival and then surpass land revenue in the second half of the nineteenth century. By the early twentieth century, revenues from just one forest product—pine resin—were more than double the entire land revenue in the hill districts of Kumaon (Trevor and Smythies 1923: 40). With the increase in their worth came assertions about the wastefulness and unsustainability of existing methods of exploitation. It did not take long for a new environmental regime to be born: defining what forests meant, how they should be managed, and for what purposes they should be used.

The Consolidation of Government, 1868–1921

Ramsay continued to serve as the conservator of forests until 1868. A new forest department was constituted in 1863 following a conference in Nainital, which all Indian conservators attended (Chaturvedi 1925: 358). Major G. Pearson became the first independent conservator of the Northwest Provinces and elaborated further the conservation measures initiated by Ramsay. The forest department began to map areas under its control, frame regular working plans, open unworked forests, and record existing rights in forests (Pearson 1869). The overall shortage of trained foresters at this time meant that most of the staff of the forest department was recruited from among engineers, military officers, and naturalists. After 1870, an influx of trained foresters changed the character of departmental activities. As these foresters assumed senior positions in the department, they increasingly began to use interventionist procedures based on calculations of growth, volume, and yield and numerical estimates of valued species of trees. Between 1870 and 1900, processes of forest making similar to those elaborated for India in chapter 2 ensued in Kumaon; these included new categories of protection, increased emphasis on timber production and marketing, direct involvement in the collection of resin and manufacture of turpentine, and refinement of regulations.

Around 1875, the forest department began to prepare working plans to regulate the felling of trees. These plans detailed the area of forest to be worked, the number of trees to be felled, and operations needed to improve the forest. A working plans division was formed in 1880, and department officials began to develop regionally specific models of forests with normal increments, distributions of age classes, and growing stock. The objective of defining normal was to make existing vegetation approach the values associated with the term (Trevor and Smythies 1923). Within about three decades, the forest department had taken over nearly half of Kumaon and had classified nearly 85 percent of the area under different forms of government based on working plans. It devised appropriate methods of planting and harvesting and applied them to specified blocks. As foresters developed more reliable quantitative estimates of the relationships among age, growth rates, diameter and girth, volume, yield, and value, they began to congratulate themselves on their success in rejuvenating vegetation in a province that they had inherited in a ruined state (Chaturvedi 1925: 366; Robertson 1942: 55).

Kumaon saw the introduction of fire protection in 1876. Regular records helped keep track of the number of fires; their location, origins, and motivations; and the degree of success in protecting forests from them. By 1901, more than 65 percent of the area under the control of the forest department was being protected from fire (Osmaston 1921). Together with fire protection, the department also sought to restrict cattle grazing, fodder harvests, and lopping. Finally, department officials classified many nontimber plant species as parasites and undertook regular operations to eliminate them. Going beyond what Brandis had attempted in Burma, Kumaon foresters began frequent creeper-cutting operations in 1887 in several working plan circles.[15]

The establishment of the Forest Research Institute in Dehradun made the United Provinces the center of new research in silviculture and forestry. By 1906, R. S. Troup had initiated systematic statistical research on several valuable timber species, including chir (*Pinus roxburghii*), deodar (*Cedrus deodara*), and sal (*Shorea robusta*). Other officials were producing comprehensive listings of Kumaon's flora (Osmaston 1927). The yield tables Troup created for sal dramatically changed the rotation, thinning, and yield calculations in later working plans. The drafting and implementation of working plans was necessarily a slow operation, but by 1924 the forest department could claim in its annual report that all the areas classified as reserved forests in the United Provinces were under either a sanctioned working plan or a plan that was already being implemented (UPFD 1925).

The Forest Act of 1878 created three new categories of protection: reserved forests, protected forests, and village forests. In reserved forests, the forest department could restrict all activities by local users, including grazing and lopping of fodder or firewood, without explicit permission. Although the state could create reserves under the 1865 act, the extent of restrictions in reserved areas was far greater under the 1878 act. Village forests were those lying within the boundaries of a village. In Kumaon, these were forests that lay within boundaries demarcated by Traill in his settlement of 1823. Villagers could have unrestricted access to their village forests. Protected forests were all uncultivated lands not explicitly classified as reserved forests or gazetted as village forest.

The area under the control of the forest department increased substantially by the 1890s. In 1893, the department converted all waste and uncultivated lands into district protected forests.[16] Two additional notifications in 1894 reserved commercially valuable trees such as chir (*Pinus roxburghii*), cypress (*Cupressus turulosa*), deodar (*Cedrus deodara*), sal (*Shorea robusta*), sissoo (*Dalbergia sissoo*), tun (*Toona ciliata*), and khair (*Acacia catechu*) and restricted the activities permitted in protected forests. Cutting timber, selling it, lopping for fodder or firewood, hunting, and the use of traps were all severely curtailed (Rawat 1991: 286–87).

But enforcing these regulations proved exceedingly difficult. Senior administrators complained that although the laws forbade the felling of trees there was no way to enforce the laws owing to the lack of trained personnel and resources.[17] According to the deputy commissioner of Almora, it was unreasonable to suppose that the department "could ever secure real control over the district forests with only one ranger to 4,832 square miles, and one forester to 371.7 square miles of forests" (quoted in Shrivastava 1996: 183). Such difficulties in protecting the district and other protected forests led finally to the massive conversion of the protected forests into reserved forests. Between 1911 and 1916, the area classified as reserves increased from 200 to more than 3,000 square miles (see figure 5)![18] The process of demarcation and actual reclassification was slowly completed by about 1920.

The department's conversion of large areas of land with or without much vegetation into reserved forests helped the cause of exclusionist protection by reducing its costs. Since villager activities in reserved

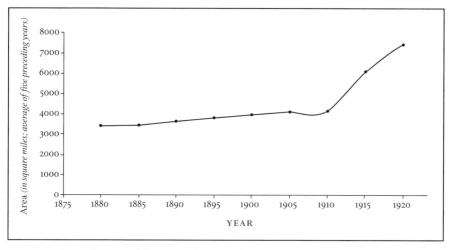

FIGURE 5: Area of reserved forests in the United Provinces, 1880–1920.

Source: Annual progress reports of the forest administration in the United Provinces, 1880–1920.

forests were prohibited, any villager found in the reserves was automatically in violation of forest laws. In district and protected forests, villagers could always protest that they were extracting fodder or firewood up to the permitted levels. The conversion into reserves was also in line with the stated policy of regulating and expanding the production of timber (Ribbentrop 1900). The department's success in preparing working plans for the areas it controlled and the revenues it procured from what it already possessed made it eager to acquire new land. Three additional reasons served as justifications for the expansion of territorial control. Department officials linked forest protection through "scientific forestry" to soil conservation and stabilization of hill slopes. They equated the felling of trees without careful regulation with unpredictable and undesirable climate change and flooding in the lower reaches of rivers originating in the mountains. And they identified local practices with deforestation and environmental destruction.[19] Through such justifications, the forest department came to defend efforts to control and regulate a vast area of land that it classified into different types of forest.

However, the territorial gains of the forest department were short-lived. Its pursuit of what it called scientific forestry ran into the obstacle of powerful opposition from the revenue department and determined protests by rural residents. To make the regulations permanent, it had to devise a new strategy and concede control over substantial areas of new reserves. The new strategy had the requisite effect: those contesting control—revenue department officials and rural residents— became willing accomplices.

Cracks in the Regulatory Edifice

Officials in the land revenue department often attacked the forest department's efforts to increase its territorial possessions. They questioned the very logic of land appropriation by foresters. The power and income of the revenue department were in part based on the amount of land it could govern and tax. It therefore argued in favor of expanding agriculture and cultivation. The forest department's claim that it had the legal right to decide the fate of areas that were not cultivated naturally became a bone of contention between the two departments.[20] Each claimed the right to administer its own tracts of land: reserved forests for the forest department and cultivated land for the revenue department. But each wanted also to control the remaining uncultivated wastelands. The strategy for the forest department was to declare these lands forests and progressively convert them into reserved forests. For the revenue department, the strategy was to question this classification by raising doubts about whether reservation was necessary for such large areas in the hills.

Like a Dog in the Manger

Tensions between the two departments were evident even in the 1860s, when the forest department was created. After 1893, as the forest department sought to extend its reach into Kumaon's territory, these tensions often surfaced. Official correspondence between the provincial government and officials in the revenue and forest departments makes these tensions and their reasons quite clear (Farooqui 1997: 52– 55). The forest department, having increased the income from forest exploitation to historical highs, was unwilling to cede authority over the forests it had created. The revenue department was similarly loath to give up control over virtually the entire territory of Kumaon to a rival. V. A. Stowell (1916), a deputy commissioner in the revenue department, protested the work of the forest department, arguing that "the Commissioner in the hills is not overworked, you are taking away his civil and judicial work if you lop the forests out of his administration. What is left for him to do? Practically nothing apart from his loss of all authority of the most vital interests of his district" (quoted in Baumann 1995: 86). Another revenue official claimed that forest department officials "apparently do not want the forests themselves, but, dog-in-the-manger like, they say nobody else shall use them" (Ross 1894, quoted in Shrivastava 1996: 146). A senior settlement officer compared the forest and revenue departments to "two kings in one city" (Pauw 1895, quoted in Shrivastava 1996: 147). Such tensions were not unique to Kumaon. About the same time as the passage of the Forest Act of 1878, one of its chief architects, B. H. Baden-Powell, summed up the frictions in Madras Presidency: "It is not the fault of the Madras [forest department] officers that the forests are undemarcated, are destroyed without check, that the reports repeat year after year the same sad story of fires, cattle-trespass, and waste: nor is it their fault that . . . hundreds and thousands of rupees are annually converted from Imperial revenue to local purposes. . . . It is the fault of the obnoxious system which places every forest officer in subjection to the Collector [of revenues], the relation being at once unsatisfactory and undefined" (1876: 198).

Although foresters felt they had good reasons to be pleased about transforming uncultivated lands into different types of forests, revenue department officials were critical of the new restrictions and protested that they created the potential for massive conflict with the peasantry. Especially vocal in expressing their dissatisfaction were senior revenue officials Henry Ramsay and D. T. Roberts and later V. A. Stowell and P. Wyndham. According to Ramsay, the restrictions on grazing, firing, and firewood collection were unnecessary, against custom, and unprofitable. His belief that the forest department was extremely unpopular was shared by those outside the revenue department. V. A. Stowell argued passionately in 1912 about the prevailing hatred of Kumaonis for the "horde of forest department underlings" and "survey parties that had flooded the hills" (quoted in Farooqui 1997: 49). Even some forest department officials admitted the unpopularity of their conservation measures (Chaturvedi 1925). It was only the reasons for unpopularity that were a matter of dispute. Revenue department officials argued that the unpopularity of forest-related measures was unnecessarily earned. Their counterparts in the forest department insisted on the foresightedness of the new laws and attributed their unpopularity to the ignorance of the peasants.[21]

Clashes of interest between the forest and the revenue departments were common in other parts of India as well. Intervening in such a dispute between district collectors and foresters in Bombay Province at the turn of the century, the secretary of state for India enjoined the need for cooperation, observing that "the proper growth and protection of forests is as important to the government as the cultivation of any other crop" (UPFD 1961, quoted in Baumann 1995: 85). But these differences were so pervasive in Kumaon that the provincial government called a meeting of the region's revenue and forest officials in 1916. The meeting was held in Nainital, and was attended by James Meston the lieutenant governor of the province, together with most senior forestry and revenue officials. The meeting settled two questions: to what extent was the reservation of almost the entire area of Kumaon to be upheld, and were revenue authorities empowered to exercise any control over the forest department? Revenue department officials considered grazing fees, the supply or sale of forest produce to villagers, budgetary proposals regarding new buildings, and assessment of the performance of the forest department to be properly within their purview. To their dismay, both questions were settled in favor of the forest department. The forest department was to continue to decide on all matters related to forests, even if they impinged on land revenue and village life, and the policy of reserving forests was to be continued (Farooqui 1997: 54–57).

The more exclusionist views of foresters in Kumaon succeeded in no small part because of the success of the forest department in producing large surpluses year after year. Revenues from forests were relatively small at the beginning of the nineteenth century. By the early years of the twentieth, they had multiplied by many orders of magnitude. India's export surplus was critical to the colonial state's balance of payments, as Indian economic historians have pointed out (Bose and Jalal 1998: 100). The forest department in Kumaon produced revenues and surpluses far in excess of those accruing from land. It was because of this success in revenue generation that even outside observers found the work of the forest department praiseworthy.[22] A cursory look at the

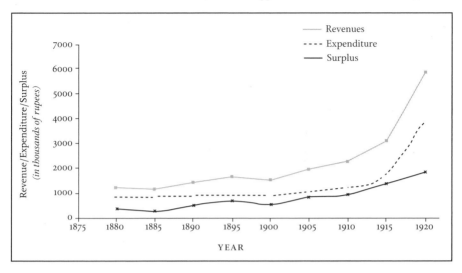

FIGURE 6: Financial statistics of the United Provinces forest department, 1880–1920.

Source: Annual progress reports of the forest administration in the United Provinces, 1880–1920.

financial performance of the forest department makes clear its commercial success, despite the professed aim of conservation.[23] From its
formation, the revenues and surpluses of the department had increased
steadily (see figure 6).[24]

High revenues did gain the forest department a victory in its efforts
to reserve a large portion of Kumaon. But this victory did not last long.
The retreat that forest department officials beat in less than a decade
did not occur simply because of the efforts of rivals in the revenue
department. Protests by Kumaon's villagers forced the issue more de-
finitively. The fires they set vitiated the basis of the successful performance of the forest department—planning and care.

The Growth of Peasant Protests

The revenues and surpluses of the forest department were directly
related to the increasing amount of land under its control and greater
restrictions on villagers' activities. They thus came at the expense of
what villagers considered to be their rights in forests: the rights to
collect firewood, graze animals, harvest fodder, and cut timber. The
restrictions naturally made the forest department extremely unpopular

throughout Kumaon (Shrivastava 1996: 190–205). As early as the end of
the nineteenth century, villagers had attacked foresters who were trying to reserve areas the villagers considered sacred groves (Rawat 1991:
308–10). Guha ([1989] 2000) and Farooqui (1997) describe at length the
wide range of strategies villagers adopted to protest the forest reservation policies of the government.

The forest policies of the government were also condemned in various parts of Kumaon by local leaders. After a large public meeting in
1907 in Almora, the local leader Badri Dutt Joshi lodged an emphatic
protest against the policies with the government (Pant 1922: 44). Even
government officials noted the depth of discontent among villagers
(Farooqui 1997: 79–80). It was not just the government policy regarding forests that led to remarkable discontent. Kumaon villages had also
been burdened with the coolie system of forced labor extraction since
the early nineteenth century.[25] Under the coolie system, villagers could
be recruited to meet any labor need of the administration.[26] Noted
leaders in the hills established the Kumaon Parishad (Kumaon Council)
in 1916. The aim of the *parishad* was to abolish forced labor (Pathak
1991).

In part as a result of these protests, the lieutenant governor of the
province, Sir John Hewitt, claimed somewhat disingenuously in a
speech delivered in 1912 at Haldwani that "the Government had no
desire to make money out of the forests in Kumaon and intend[ed] to
spend for the benefit of the people of Kumaon, the amount by which
the receipts exceed the expenses."[27] The existing orders could not be
changed, however, because they were necessary for the protection of
wooded areas from "reckless destruction" (quoted in Pant 1922: 44). Sir
John had promised Kumaon's villagers in 1912 that there would be three
classes of forests: Class A would consist of areas intended principally
for protection, Class B would contain fuel and fodder reserves, and
Class C would be freed from official control so that people could exercise usufruct rights in them.[28] But by 1917 the rights people enjoyed in
class B forests had been eliminated altogether and many of the trees in
class C forests had become reserved. The new reservations helped increase simultaneously revenues for the forest department (Gibson
1920: 360–61) and the ire of villagers against it.

Especially irksome to villagers was the conversion of almost all uncultivated (*benap*) land into protected forests in 1893 and then their

gradual reclassification as reserved forests between 1911 and 1920. This change left villagers almost no uncultivated land outside reserved forests. By 1917, the cultivated area of Nainital and Almora was less than 10 percent of the area of reserved forests. Class C forests, under the control of the revenue department and from which villagers could collect forest products for their needs, amounted to only about 100 square miles in Nainital (Pant 1922: 53).

In addition to prescribing limits on the geographical area in which villagers could enter and use forests, the forest department also specified elaborate new rules to restrict lopping and grazing rights, reduce the extraction of nontimber forest products, prohibit the extension of cultivation, and enhance the labor extracted from villagers. It hired more guards to implement and enforce the new rules. The rules were complicated, and enforcement was not as strict as the letter of the law required. But the increase in the size of the forest bureaucracy still created opportunities for guards to extort bribes and raised the level of hostile interactions with village women and children who harvested products from the forest.

Friction with forest guards occurred in part because they were enforcing a stricter set of regulations over a larger area classified as reserved forest. It was also inevitable because forests were a critical part of Kumaon's agricultural economy. As Pauw, the settlement officer, said, the people in the hills were "no less pastoral than agricultural" (1896: paras. 24–25). Animals provided dairy products, manure to fertilize agricultural fields, and draft power for tilling. Agricultural productivity was an almost direct function of the amount of manure added to the fields. Forests were the source of firewood for heating and cooking. Thus, rural residents depended on forests for fodder for livestock, manure for fields, and firewood for the household. Links between forests and agriculture continue to be critical to this day.[29] Even Baden-Powell, indefatigable in his defense of state ownership of all forests, conceded that villagers in Kumaon might have some rights in forests precisely because of the close connection between agriculture and forestry.[30]

Although government officials asserted that the new laws were never strictly enforced, they goaded villagers into "violent and sustained opposition" (Stebbing 1922–26: 258).[31] The annual administration report of the forest department in 1916 was explicit. "In the Kumaun circle," it noted, "there was an epidemic of fires in the whole area: 44 are ascribed to organized incendiarism. . . . This appears to be proved by the fact that numerous fires broke out simultaneously over large areas and often the occurrence of a fire was the signal for general firing in the whole neighbourhood. . . . The resulting damage was enormous" (GOUP 1916: 7). According to one forest officer (Chaturvedi 1925: 366), "the formation of the 'reserves' led to an epidemic of forest offences which culminated in organized incendiarism in 1921."[32] Villagers simply refused to accept the rules. Nor were they willing to concede the fundamental assumption undergirding the new rules: the state's power to constitute a monopoly over all natural resources it deemed significant.[33] The best efforts of forest department officials failed to convince the villagers that forests belonged to the state. The officers who had designed the new settlement had hoped that the residents of the hills "would gradually become accustomed to the rules as gazetted and that control may be tightened as years go on" (KFGC 1921: 2). But hill dwellers dashed these sanguine hopes.[34]

Villagers broke almost all the rules they were supposed to follow, swelling the records the forest department maintained on breaches of forest rules. Many of their actions were at an individual level, oriented toward extracting forest products such as fodder and fuelwood and grazing livestock. But their offenses also reveal an interesting pattern overall (see figure 1). Prior to 1916, the number of people convicted for each detected infraction was less than two. Between 1917 and 1921, the number of people convicted for each infraction rose to somewhere between five and six. After 1926, the average dropped down again to less than two (Guha [1989] 2000; Agrawal 2001a). Collectively organized breaches of forest law thus occurred far more often at the peak of the new restrictions.[35] Although the direct testimony of villagers is not available, it is clear that their dissatisfaction and responses were expressed far more collectively in the period 1917–21 than before or after it. In light of this, it is hard to accept the official suggestion that Kumaonis "never half understood the rules that were made and often had vague ideas of the entries within their rights lists" (KFGC 1921: 2). Villagers may not have known the exact text of the rules, but they understood the implications well enough. They acted collectively against the rules with all that such actions imply in terms of joint discussion and understanding and the reimagining of personal interest. Indeed, villager protests were so vehement that the annual report of the forest department for 1922 accepted that "owing to the anti-forest campaign, the machinery for the control of forest offences more or less broke down." Grazing infractions increased so greatly that forest officials found "the pressure of grazing combined with the lawless attitude of the people a serious menace to the tree forests" and in some cases found the position so bad that they contemplated the "abandonment of the forest" (GOUP 1922: 7).

Villager protests against forest reservations intersected with agrarian unrest and their grievances against the system of forced labor extraction in powerful ways (Siddiqi 1978). Throughout the period 1916–21, there were massive demonstrations in Kumaon. Demonstrations and strikes led the forest department to abandon many working plans. Its 1921 annual report admits that "owing to the [coolie] utar strike and the general political conditions in Kumaun preparations of the East Almora working plan had to be abandoned for the present which is much regretted as forest management in this division is far from satisfactory. The whole subject of forest policy in Kumaun is now under review" (GOUP 1921: 2). The report also warned that if the forest department did not receive the authority to close off forest areas and take appropriate measures against fires "no scientific management" would be possible (2).

In their demonstrations and meetings, Kumaon residents showed their unremitting opposition to the coolie system. The movement took a turn toward greater radicalism in 1921 with a series of continuous meetings in different Kumaon villages. Hundreds of villagers courted arrest, and many villagers simply refused to perform the labor required of them (Pathak 1991: 270–76). Protests against forced labor became especially fierce that year, and incendiary fires in forests registered their greatest upsurge. Another critical development was the return of nearly ten thousand Kumaonis whom the British had recruited to serve as soldiers during World War I (Farooqui 1997: 80). The fear that these soldiers, with their deep local ties and experience of armed fighting, could set the whole of Kumaon ablaze contributed in no small way to the official willingness to listen.

"The Strictness of Popular Control":
Community and Conservation in Kumaon, 1921–1931

The incessant, often violent, protests against reservations occurred in the context of tensions and rivalries within the colonial state between its two most powerful arms in the hills—the revenue and forest departments. This was coupled with discontent and visible protests against forced labor and the presence of a large number of young Kumaonis who had served as soldiers in World War I. In response to the protests, the government appointed the Kumaon Forest Grievances Committee to look into local "disaffection." Comprising government officials and local political leaders, the committee examined more than five thousand witnesses from all parts of Kumaon.[36] It used the resulting evidence to make nearly thirty recommendations. The committee felt that many of the breaches of law by villagers were simply a result of unenforceable rules that interfered directly with the actions of villagers aimed at securing a livelihood. It advocated the repeal of all restrictions on grazing and lopping of fodder from oak trees.

The committee also recognized the social power dynamics surrounding the enforcement of the new laws when an agent of the state such as a forest guard was empowered to cite villagers for actions that constituted for them no more than everyday use of the forests but were seen as infractions under the law.[37] In such a situation, not only were a larger number of guards needed to enforce the law but enforcement itself promoted dissatisfaction. In addition, the law created opportunities for guards to extract bribes even for minor infractions. The committee suggested that forest department employees would be prevented from harassing villagers, especially where women and children were involved, if they guarded a smaller area of forest and were kept busy with other departmental work. The committee's observations about limiting the area of forest under the control of the forest department matched the department's own earlier complaints about not having enough employees to guard its possessions. But the committee's solution differed from that of the department. Instead of advocating an increase in the number of employees, it suggested a reduction in the area of forests!

The most significant concrete suggestions of the committee were twofold: (1) reclassify the majority of the reserved forests created between 1911 and 1917, and (2) lay the foundations for creating community forests that would be managed under a broad set of rules framed by the government but for which villagers themselves would craft specific rules for everyday use. The government took both recommendations seriously, much to the dismay of forest department officials. The 1922 report of the forest administration argued that the reservations had resulted in far better protection of forests, the extension of the resin and turpentine industries, and higher revenues.[38] The 1925 report complained that the committee had given to villagers "virtually everything they asked for in the way of unlimited grazing and forest produce . . . [but] the case of Kumaon was one where a population ought to be protected against itself, however unpopular such action may be" (Anonymous 1926: 47–48)!

At first, the reserved forests that had been taken over by the forest department between 1911 and 1917 were reclassified as class I and class II forests. Class I reserved forests were of relatively limited commercial value. They contained broadleaved tree species on which villagers relied for fodder and fuel. Class I forests also comprised smaller patches of vegetation (less than 1 or 2 square miles) located close to the village. All of these forests were transferred to the revenue department as district forests and in time could come to be controlled by villagers if they followed a specific procedure described in the 1931 Forest Panchayat Rules. Class II reserved forests contained large contiguous areas of commercially valuable species such as chir, sal, deodar, and cypress. The forest department retained control over them. But even in the class II forests villagers gained the right to collect dry twigs for firewood and cut grass for fodder.

By 1927, nearly 2,000 square miles of class I forests had come back under the control of the revenue department. Another 1,300 square miles of class II forests remained with the forest department, of which 200 were the old reserves dating from the 1890s. Over the next sixty years, a large proportion of the class I forests and even some of the reserved forests came to be managed by village-level forest councils (van panchayats) as community forests.

At least initially, the formal transfer of 2,000 square miles of forests to the revenue department by 1926, and, more radically, to village councils over time, was seen by forest department officials as a move toward their destruction "in view of the shortsighted outlook of the inhabitants."[39] The annual report of the forest department for 1925–26 warned on its very first page that the forests that had been transferred were unlikely to survive and "their ultimate destruction is probable. There are proposals to manage some of them by village panchayats, which may retard their disappearance, but in the opinion of the writer will not save them to posterity" (GOUP 1926: 1). Over the next decade, forest officials were repeatedly to predict the dire consequences of the change in the administration of forest resources.

For the forest department, the creation of village-level council forests and the transfer of land to them were bad enough. Even more difficult to swallow was the bitter pill of knowing that these forests and forest councils would be under the control of the revenue department. After the provincial government had accepted the recommendations of the Forest Grievances Committee, foresters complained: "The draft rules [for the council forests] submitted to the government for approval leave control entirely in the hands of the Commissioner and the Deputy Commissioner while the Forest Officer is merely to be consulted, when necessary, in the capacity of advisor" (Anonymous 1931: 240). The annual report of the forest department stated: "The work of the forest department in regard to panchayat forests is limited to giving help and technical advice" (UPFD 1930: 2). It was clear that popular mobilization, villager protests, revenue department hostility, and the use of forests as a political issue in local elections had made it impossible for forest conservancy to be only a technical or scientific subject.[40]

Conclusion

The forest department in Kumaon remade the meaning of forests and institutionalized their care and maintenance by the end of the nineteenth century. The rules and forms of governance it endorsed and attempted to implement were parallel to those pursued in other parts of India. Protection from fire, exclusion of local residents and their animals, and systematic planting and harvesting, all based on careful classification of land into administrative units and their planned exploitation through working plans, were the highlights of normal forest department activities. Calculation and numbers played a major role in the production of the idea of forests.

In developing the above themes in relation to Kumaon's landscape, this chapter follows the lead of chapter 2. But it also shows that forest department officials did not always realize their plans. Although the forest department was victorious in its initial skirmishes, the close links between hill agriculture and forests meant that the actions of the department in Kumaon had far more adverse impacts on local residents than did the actions of its sister departments in other provinces of India. Territorial aggrandizement and restrictions on rural residents' actions in forests resulted in a series of protests between 1916 and 1921. These protests coincided with resistance against forced labor and the return of more than ten thousand Kumaoni soldiers. The presence of widespread grievances coupled with tensions within the colonial administrative apparatus led the state to concede the demands of the Kumaonis.

A number of measures between 1921 and 1931 concerned with the disposition of almost 2,000 square miles of forests finally found form in the Forest Council Rules of 1931. These rules permitted village residents to create forest councils and control the class I reserved and civil forests that were under the control of the revenue department. The historical process by which forests came to be regulated collectively (though not necessarily without tensions) by the forest department, the revenue department, and village residents is analogous to the policies of the British colonial state in some of its other possessions (Shrivastava 1996: 301–2) and in the Madras Presidency in India as well.[41] As early as 1863, Dietrich Brandis contemplated a category of village forests from which villagers could meet their needs for firewood, fodder, and nutrients for agricultural fields. It was laid down as ultimate policy even in the United Provinces in 1912 (Mobbs 1929: 470). Baden-Powell wrote in 1892: "The day will probably come in India when village bodies will hold regular forest estates . . . [and] in time we may hope to see villages or groups of villages regularly owning well-managed forests" (203, 237).

Foreseeing the continuing formation of forest councils and development of a "movement" in favor of local protection, the chief conservator of forests in the United Provinces stated in 1930: "My own impression is that if popular control of forests is shortly attained, Kumaun may first obtain some relaxation, but will before long be called on to pay its way and that *popular control may in the end insist on much stricter control of Kumaun than we can at present enforce*" (Canning, quoted in Srivastava 1996: 296, emphasis added). Canning's faith in the ability of local populations to protect forests hinged on the unarticulated assumption that they, too, would come to see forests as did the forest department: exhaustible resources that needed care. The transformation of forests into calculable resources and of Kumaon's residents into environmental subjects is the process that the chapters in the next part of the book investigate.

PART II

overleaf: One of the uses of Kumaon's forests
Photo by Arun Agrawal

II

A New Technology of Environmental

Government: Politics, Institutions,

and Subjectivities

The first part of this book explored the technologies of environmental government implicated in the making of Indian forests. In looking at different strategies of power/knowledge, it paid special attention to the role of numbers and statistics. Centralized government of forests depended on the use of numbers and numericized relationships to represent forests, exercise control, and govern at a distance. It used the example of Kumaon to show how plans to govern seldom translate into practice as intended. In Kumaon, the most obdurate obstacle to exclusionist government was the opposition of the region's residents, who relied on the same forests for a significant portion of their livelihoods.

This part of the book examines the reconfiguration of a new technology of government that relied on localities as partners for successful regulation. The next three chapters focus in turn on three crucial aspects of decentralized environmental regulation: (1) the redefinition of political and administrative links between the state and localities, (2) the realignment of institutional and social relationships within local communities, and (3) the emergence of a more widespread concern with the environment and the making of environmental subjects. It is worth pointing out that, although initially I use the terms *locality*, *state*, and *community* seemingly uncritically, part of the objective of the discussion is to show the constructed nature of each of these terms and the usefulness of the concept of technologies of government for rethinking them (Abrams 1988; Agrawal and Gibson 1999; Mitchell 1991; Moore 1996b; Raffles 1998).

Chapter 4 examines the first facet of decentralized environmental regulation by focusing on what I call the governmentalized locality. Dispersal of power is an obvious characteristic of decentralized government. Instead of an identifiable point and marker, a single source and logic that the early efforts of the British represented in Kumaon, power over forests is now scattered and emanates from multiple locations.[1] With the increasing number of forest councils, no longer can a single agency such as the forest department be identified as the repository of power. Instead, in any given locality, a multiplicity of strategies, agencies, and forms of power are visible. The exercise of power is highly modulated to variations across settings: in vegetation, social landscapes, productivity levels, articulation with market forces, connections to other centers of power, and so forth. The proliferation of sources of power is unavoidable because the usual binary of domination and resistance can no longer be mapped onto the forest bureaucracy and the locality, respectively. The locality itself is divided—most obviously against itself—as multiple agencies and forms of power compete over claims to forests.

This suggests a second obvious characteristic of decentralized government. Within each territorialized, governmentalized locality, many different agents are involved in regulation: the headman, the council of elected representatives, the guards appointed by the councils, and, depending on forms of monitoring and enforcement, the residents themselves. Together with the increase in the number of agencies of power, there is also a proliferation of the strategies, flows, and directionalities of power. Headmen, guards, and council members are officials, but they are also local residents who meet other villagers frequently and whose positions depend on villagers' support. They are involved in complex relations of sociality and reciprocity that are only inadequately described by unidirectional mappings of domination and resistance. Governmentalized localities constitute an "effort to adjust the mechanisms of power that frame the everyday lives of individuals; an adaptation for and a refinement of the machinery that assumes responsibility for and places under surveillance their everyday behavior, their identity, their activity, their apparently unimportant gestures" (Foucault [1975] 1979: 78). In contrast to the environmental control that the forest department sought to enact, the government of environment crafted and implemented by the forest councils touch the lives of their targets far more lightly, regularly, intimately, and in proportion to their activities in the forest.

The increase in the loci through which regulation is effected implies a concurrent change in the nature and prospects for collective action. No longer is the state or even the forest department the center of the governmental power that shapes practices in forests. No longer are large masses of humans ranked against singular injustices. Instead there is a fragmentation of government and a multiplication of localized efforts to shape it. Within the governmentalized locality, the more familiar means of altering the way forests are regulated is likely the individual defense of livelihoods, not a collective movement to revolutionize institutional power. Since villagers can express their complaints through formal institutional mechanisms, they are less likely to take to the streets or set fires in forests. The transformation of the relationships between the center and the locality may occur in a manner that slowly dissolves the differences through which the boundaries between the state and the community are policed. But the dissolution of these boundaries occurs together with a revision in the relationships between the state and its subjects.

The proliferation of the nodes, forms, modalities, flows, strategies, agencies, and practices of power does not mean that such multiplication is chaotic or unruly. Instead the articulation between centralized and localized sources of power occurs according to a new combinatorial and sequential logic. Localities articulate with the state on a one-to-one basis and with each other only minimally or not at all. When they do interact, their relations are often conflict laden, as when residents of one settlement extract products from the forests of another. New procedures of rule generate administrative channels that facilitate highly asymmetric flows of power. Take, for example, the written records that forest councils maintain. These records make information about the behavior of council members, the condition of the forest, infractions of environmental rules, illegal activities in the forest, and council incomes and expenses available to state officials and other powerful outsiders. But forest council members themselves have little access to administrative processes within the provincial state machinery.

Further, localities enter into formal institutional relationships with the state piecemeal. Instead of the forest department extending its control over vast areas of forests en masse, a process that is vulnerable to obstacles and resistance, control is extended locality by locality. The state gives up the idea of protecting and improving all the forests by a single stroke of the administrative pen. But it concedes the grand project of extensive control only in return for a surer means of intimate regulation. The Forest Council Rules help the decentralizing state determine the spaces of tolerable illegality. Decision makers in the locality define the depth and nature of regulation within that space. State officials participate in what might be called flexible regulation, which becomes more entrenched for all its flexibility. The greater ability to regulate everyday practices on the part of the locality becomes the coin with which the state buys relief from threats of fire, arson, and other violations of forest laws.

More precise regulation is ensured as well by the extension of government on this finer scale. Localities become formally affiliated with the state as their members express an interest in pursuing such relationships of environment-related regulation. There is a curious marriage, then, of interest and regulation. To illustrate, a forest council is formed in Kumaon after a third of the village residents express an interest in creating one. This implies that those groups in which a significant proportion of the population has a strong interest in creating institutional connections with the state are the ones that first become part of a network of partnerships. State officials encourage local residents and leaders to affiliate themselves with the regulatory net by pointing out the primary benefit: control over the allocation of products from the forests they will come to govern. Such education of the local leadership began as early as the 1930s in Kumaon, immediately after the passage of the Forest Council Rules. Through material rewards and knowledge transfers the state manufactured a new interest in the governmentalization of the environment and the locality (chapter 4).

Chapter 5 directs attention to the processes embodied in the regulatory community that comes into being side by side with the governmentalized locality. The shift in the exercise of power from the centralized state to the decentralized locality is not just about the creation of multiple locations of power and the creation of governmentalized localities. Nor is it only about the rhetoric of empowerment—of the community and its members.[2] It is also just as much about a new economy of the power to regulate, a better distribution of regulation, and a more modulated application of power to shape individual practices and subjectivities. In contrast to the governmentalized locality that signifies a new regime of political relationships between the state and the periphery, the regulatory community denotes a redefinition of relationships among different groups within the community.

In choosing the concept of the regulatory community to analyze the government of environment, I depart from Foucault's ([1975] 1979) two preferred metaphors, which describe the potential mechanisms through which disciplinary power works: the coercive institution and the punitive city. The model of the coercive institution that according to Foucault came to colonize almost the entirety of penality is Bentham's well-known panopticon.[3] The punitive city would have served disciplinary power equally well, but it was never adopted as a deterrent.[4] Neither of these models turns out to be very relevant to environmental government through the community. It would be fair to suggest that a well-functioning regulatory community obviates the need for other forms of penality and does so in a manner that appears to be far more humane. Communal regulatory authority does not need the loud and spectacular displays of retribution that sovereignty requires. Nor does it need the constant and all pervasive hammering home of representations that establish an irrefutable connection between crime and the return effects of crime upon the criminal. And certainly it does not need the crutch of the all encompassing physical gaze as a means of ensuring compliance. Instead it relies on intimate knowledge about each member and deploys this knowledge through a patchy system of monitoring and enforcement that limits infractions effectively by bounding them within a sphere of tolerability.

In this regard, the initial important accomplishment of the regulatory community in Kumaon was to force an institutional and social split among community members. In the early part of the century, community leaders and members were aligned together against the forest department's regulations. But their interests suffered a division once communities began to regulate forests as members of a network created by the state. The community came to be the agent of decentralized environmental regulation.

Today decision makers within the community use their intimate knowledge about members of the community to ensure that power is wielded neither too forcefully nor too weakly. They don't want to provoke protests or be rendered ineffective. Community regulation operates more constantly, consistently, effectively, and transformatively on its objects: village residents. Regulation is more comprehensive but less costly, more modulated but less visible, more autonomous but more continuous, more precise, and, perhaps for that reason, more humane.

Indeed, new strategies of regulation through decentralized institutions could scarcely remain in place without the greater efficiency they permit or the savings they allow the state to effect.[5] The effectiveness of the regulatory community is evidenced in the reduction of economic, social, and political costs. The fiscal burden of the state is lowered in the first instance by the reduction of the number of administrative and enforcement personnel it requires. There is no need to devise uniform criteria of selection, apply them to select guards, train the selected individuals, create mechanisms for their supervision, and pay for the costs of these procedures at scales of remuneration that are at a rough parity with salaries and costs incurred in other government departments. The regulatory community bears these economic costs. It governs forests, devising rules to use and manage them, planting trees and helping to harvest them, allocating fodder and firewood in proportions necessary to meet household needs, appointing guards and paying them a salary that is far lower than those paid to centrally employed guards, sanctioning rule violators, and settling disputes. It performs all of these activities at a fraction of the cost that the forest department would incur.

The reduction in economic costs is accompanied by similar shifts in the nature and levels of political and social costs. By transferring the tasks of protection and enforcement to the regulatory community, the central state no longer need bear the resentment or ire of those dependent on forests. Since the forest department and its guards are no longer responsible for translating into practice the rules necessary to exclude villagers from forests, the state is no longer seen as the agency of exclusion. By changing the nature of protection and the structure of authority relations through which protection is made manifest, the regulatory community also makes redundant the frustration and anger over bribes and corruption in which a government-appointed forest guard is inevitably implicated. Thus, the new strategy of dispersed regulation replaces an excess of erratic and expensive enforcement with an economy of comprehensive and continuous obligation.[6] It installs a whole network of environmental relations that, compared to earlier, more repressive forms of government practiced through the forest department, are more economical and effective, more dense and widespread, more autonomous and enveloping.

The transfer of the design and enforcement of regulations down to

the lowest stratum of a social organization can be successfully maintained only by a simultaneous transformation of the relationship between the enforcer and the offender. When the agent of enforcement is
an employee of the forest department, the offender who violates rules
and procedures needs either to be excluded from the forest or to make
monetary recompense. For some offenders, the level of punishment
exceeds the nature of the crime. Conversely, other offenders escape
punishment altogether. All those who are not part of government (and
even some who are) potentially belong to the class of offenders. The
objective of the state is to inoculate forests from future illegal actions.
Not surprisingly, one of the primary means of ensuring the safety of the
forests is to ensure the exclusion and expulsion of the offender from the
forest. But when the task of enforcement is in the hands of the regulatory community offenders are often within the community or situated
as neighbors. Their actions are the source of a disharmony that requires
balance. Offenders cannot be excluded, and, even if in exceptional
circumstances they are, their families and relatives continue to be a part
of the community. Rather than a retaliatory measure, punishment becomes a means of correcting behavior. The need to be precise in imposing punishment does not result from the likelihood of violent reactions.
Rather, it is born out of the recognition that group members who are
punished will remain members.

As a result, there is a more precise calibration of the allocation of
benefits from forests, the monitoring mechanisms that are deployed,
the type of sanctions that are imposed, and the nature of dispute resolution that is available to settle conflicts. The greater ability of the
community to monitor a wide range of practices in the forest (though,
of course, not all actions) results from the many monitoring mechanisms it can deploy. It can appoint a guard and structure his remuneration in multiple ways. Besides appointing one or more guards, the
community can also pursue more decentralized methods of monitoring. Indeed, many of the monitoring mechanisms derive part of their
power from the intimate, daily, multistranded contact and contiguous
residence of monitors and those being monitored. When different
community members rotate through the position of a guard, then
guards have significant knowledge about each other even without ex-
pending too much effort specifically overseeing neighbors. Similarly,
community regulation contains within itself the seeds of reciprocal

control by those who are regulated.[7] Mechanisms of accountability can colonize practices of regulation when the election of officials, appointment of guards, and their remuneration depend on contributions by community members. In contrast, protection of forests by the forest department depends ultimately on the presence of guards whose actions very often become arbitrary in the absence of further supervisory procedures. Of course, their second-order supervision from above runs the same risks of arbitrary implementation in an infinite regress (Elster 1989).

In fact, it is not just in the deployment of monitoring mechanisms that the regulatory community draws a fine line between the permitted and the prohibited. The more precisely attuned ability to monitor and detect violations exists together with more finely calibrated sanctions.[8] The blunt instruments of deliberate disregard for minor offenses and fines and imprisonment for actions deemed more egregious were the recourse of the forest department for virtually all violations committed by villagers. But the range of instruments available to the community is vaster, and the potential subtlety in how each of them can be deployed is greater. Sanctions encompass varying levels of exclusion and castigation, public reproof and chastisement, finely graded monetary penalties, imposition of socially desired tasks as burdens, impounding of property, and the threat of invoking central authority, which can often be more effective than its actual invocation. A more meticulous and thorough understanding of such mechanisms of regulation is the objective of emerging new sciences of community and the environment (chapter 5).[9]

It is also necessary to point out that a critical part of the transformed relationship between the community and its residents, and between community residents and the environment, is the change in the object of regulation. Instead of ensuring just strict protection of forests to enhance state revenues, regulation now also becomes a means of pacification, ensuring subsistence, and addressing poverty. More careful attention to the multiplicity of forest products allows government to undertake a more precise distribution of different types of forests to their most valued uses. The narrow goal of greater efficiency in surplus generation is broadened to include the operations of the community and the satisfaction of the needs of community members. The community, reciprocally, becomes concerned with how best to use, manage, and govern forests for its members.

Chapter 6 examines the making of environmental subjects. Ultimately, transformations in the relationships between the governmentalized locality and the state, the regulatory community and its members, and rural residents and their forests are linked to the emergence of new environmental subjects.[10] Indeed, the most interesting question in the two-century-long process through which forests came into being may be not about the governmentalization of social formations, nor about the production of regulatory communities, but about the creation of environmental subjects. It is critical to understand and explain how people came to accept the importance of environmental regulation, to respect the authority of the community to sanction actions that do not respect regulation, and to participate actively in regulating the behavior of their fellow community members.

Not all who are subjected to environmental government become environmental subjects. Therefore, it is necessary to tease out the mechanisms involved in the variable production of subjectivities. One can do so by building on the contributions of some recent accounts of environmental and developmental politics (Brosius 1999a, 1999b; Ferguson [1990] 1994; Escobar 1995; Li 1999).[11] More explicit attention to the processes involved in the variable production of subjects through strategies of government helps throw greater light on what remains an underinvestigated puzzle—both theoretically and empirically. It is precisely to this relationship between the disciplined production of the subject and the care of the population that Foucault points in his discussion of biopower "the fostering of life and the growth and care of population . . . [are] a central concern of the state, articulated in the art of government" (Rabinow 1984: 17). Foucault's ideas about power are as much about the procedures through which an individual body is regulated as they are about calculations for the population. Without the subjection of individual bodies to discipline and technologies of government, the object of government—whether development or environmental conservation—is likely to be stymied (see Foucault [1978] 1991: 98–99).

Surely many different forces have conspired to change the way people imagine their relationship with forests and the environment, among them experiences of scarcities, media accounts, and processes related to the governmentalization of the environment. Among the most critical of these forces, I suggest, are institutional changes in the regulatory strategies within communities and the related environmental scarcities they force people to confront. Rearrangements and transformations of institutions have important effects in framing people's interests, the way they act in relation to their interests, their involvement in the enactment of regulation, and the manner of their engagement with the processes of government. The emergence of forest councils, the mechanisms to allocate and enforce that the forest councils have constructed, and the variable participation of people in these mechanisms of enforcement each affects how environmental subjects have come into being in Kumaon.

It should be clear that new strategies to govern forests—to allocate, monitor, sanction, enforce, and adjudicate—do not simply constrain the actions of existing sovereign subjects. Instead, it is important to recognize how these strategies and their effects on flows of power shape subjects, their interests, and their agency.[12] By focusing on these strategies as the means through which individuals make themselves into certain kinds of subjects, it becomes possible to specify the micro-mechanisms at work in the reconfigurations of subjectivities. That is to say, explanations of why and when people respond to new strategies of power in particular and differentiated ways require attention to their structural locations, the extent to which they are privileged or marginalized by new strategies of power, and, most importantly, the manner in which they become implicated in the operations of these strategies. To insist on variations in the way subject positions change is also to insist on the evident fact that the effects of new forms of regulation are neither totalizing nor permanent.

To portray the transformation of people in Kumaon as the result of some epic battle between dominant forms of state power and resistant local populations would be to ignore the extent to which people and their communities are the product of power and its institutionalized exercise. To this extent, the separation that Foucault effects between institutions and the "social nexus," or between institutions and "social networks" (1982: 222, 224), takes for granted a certain fixity of institutional arrangements that disappears when one closely examines institutions. After all, institutions are as much about expectations about the future as they are about the mechanisms that prompt expectations.

Institutional analyses in political science have often been accused of underplaying the importance of politics (cf. Knight 1992). But they are especially deficient when it comes to understanding how power affects subject formation through institutional change. Even institutionalists who attend to politics often view power and institutions as the external limit on the expressions that internal processes within the subject would otherwise (in the absence of power) generate.[13] Institutions exist and develop independent of the subject, and the nature of the subject is typically an implicit shadow that lurks between the lines of institutionalist arguments. The writings of most major social thinkers prior to the disciplining of the human and social sciences—consider Marx, Durkheim, and Weber—not only theorized about social change but also advanced an argument about the nature of the subject and, more importantly, about the relationship between the emergence and organization of new social forms and the transformations in human mental lives and ideas (Rose 1999). Given the intimate connection between policy, institutions, and the self, the question of how human subjectivities emerge in arenas made by government can certainly be addressed more explicitly in political-institutional analyses. Whatever the many differences that characterize the work of scholars such as Coase and Weingast, Selznick and Simon, and Keohane and Krasner, their limited attention to the relationship between the subject and the social/governmental is common. *Preferences*, the term that relates most closely to the idea of the subject in these works, come from outside—structures of social experiences, prior socialization, and social location. But what distinguishes how they come to colonize particular subjectivities can fruitfully be explored further (Satz and Ferejohn 1994).

When it is long-term historical change in institutions, politics, and subjectivities that requires explanation, new insights emerge by conjoining modular accounts of institutional, political, and subjectival changes. It is necessary, however, to revise the conception of institutions and power as constraints and pregiven subjects (and preferences) as the material under constraint. In Kumaon, variations in people's practices and perceptions about forests, and changes in them that occurred *after* institutional transformations, suggest that we need to think of power as something more than a simple limit or constraint on thoughts, words, and actions. The beginnings of such a view of power are implicit even in Lukes's (1974) discussion of the third face of power, although he interprets the effects of power primarily in a negative sense as falling on preformed subjects. But it is the constitution of the subject itself that is in question when we try to understand social and governmental changes occurring over long time periods. New strategies of regulation in communities seldom function only to secure the cooperation of pregiven subjects or to prompt them to engage in new actions. Instead, regulations flow along channels laid through the body of the community and constitute new, variable understandings of what makes environmental subjects and how the interests of such subjects are to be conceptualized. The manufacture of interest and the redefinition of subjectivities play a key role in the construction of fresh beliefs about what kinds of practices are most attractive. Their effects work alongside those of new regulations.

4

Governmentalized Localities:

The Dispersal of Regulation

Cooperation and protection are the most important things in improving the performance of the forest councils. We have handed over many plantations to forest councils. First they did not believe that they really owned the forest. Now they know that the land is under their control; that they have to protect the forests as best as they can. The government can't guard all the forests that the councils are supposed to manage. —District Magistrate, Pithoragarh district, 1993

Well, my biggest problem is lack of transport. I am supposed to inspect more than 200 councils in a year and travel more than twenty days each month. How can anyone do that, I ask you? I only travel about seven to ten days. With about eighty days of touring time, it is not possible to visit the 200 forest councils under my control. And if we don't visit, it is not possible to know how they are doing, what we need to correct, and where there are problems.—Forest Council Inspector, Almora, 1993

We receive many petitions from forest councils—about tree plantation, villagers who are breaking rules repeatedly, and encroachments on council-managed forests. We don't take long to address them. We ask the patwari to take care of the matter. If the council wants the matter resolved immediately, sometimes that is not possible. We have many other duties as well.—Subdivisional Magistrate, Pithoragarh district, 1993

Vociferous protests by Kumaon's villagers had forced state officials to realize the high cost of centralized regulation. But the same factors that made villagers' protests so effective could also be used imaginatively to make decentralized government a success. Village settlements were highly dispersed. Their dispersal meant that villagers enjoyed far better access to scattered patches of vegetation than did government officials. This also meant that they possessed a greater capacity to regulate use. Villagers were highly dependent on fodder, green manure, firewood, and wood. Collectively, then, they had an interest in protection. Given the way existing state policies had cast villagers and their communities into an adversarial role, the incorporation of localities into processes of regulation could not be undertaken lightly or accomplished easily. Yet villagers could potentially monitor, guard, and protect the forest far more cost effectively than could the forest department. A new calculus of gains and losses, costs and benefits, and advantage and interest was necessary to transform rebellious hill men into allies of conservation and commerce.

Conservation and commerce existed in an uneasy relationship. On the one hand, it could be claimed that these two goals would always be in tension. After all, the pursuit of higher levels of profits would necessarily imply the harvesting of that which had been conserved. But there was another side to the picture. Followers of today's wise use movement would have recognized a kindred spirit in nineteenth-century forest officials in Kumaon. For the forest department, conservation was the first step toward higher timber yields and only a way station on the path to higher profits. Localities could play a role in this vision by helping conserve the forest, but their efforts at conservation had to assist the goal of profitable exploitation of Kumaon's forests. If procuring the cooperation of Kumaonis meant turning over complete control of the forests to them, the whole purpose behind the new alliance would have been vitiated. Many advocates of community-based conservation today prescribe a similar role for local communities. If local actors do not conform to externally desired goals of environmental protection, many conservationists would see little gain in creating partnerships with localities.[1]

To resolve this difficult conflict, colonial state officials chose a territorial-institutional solution. The alliance between different departments of the government and Kumaon's localized communities hinged on a division of forests. According to this alliance, villagers were given responsibility for conserving their forests in exchange for harvesting subsistence products and the forest department arrogated to itself the benefits of commerce from its own as well as the villagers' forests. To compensate villagers for their added responsibilities, the colonial state gave them back a part of what it had taken: some rights to products from forests that were used within the household to meet agricultural and subsistence needs. As one of the forest officers, J. S. Campbell

(1924) suggested to the secretary of the provincial governor, "with a modification of the rules aimed more at regulating rights than curtailing them, I think we can get the more sensible people at least to view our policy with modified approval" (quoted in Baumann 1995: 84).

The territorial-institutional division of forests went hand in hand with a new technology of government. The main aim of this regulatory apparatus was to make communities an effective junior partner of the forest and revenue departments. Regulation was founded on new legal measures, fresh administrative and ecological classifications of landscapes, additional official positions and budgets, further responsibilities for existing officials, innovative techniques to redistribute the objectives and instruments of forest governance, wider dispersal of authority and revenue allocations, and even a different way to define community. In short, localities and their residents became participants in conservation on the basis of a transformation in the understanding of the place of forests in social life.

Only over time did the different partners in this alliance—localities, the forest department, and the revenue department—come to appreciate the potential and capabilities of the new technology of government and their role in it. The most important instrument that colonial administrators used to shape relations among localities, rural residents, and forests was the Forest Council Rules of 1931. This set of rules tried to define, situate, and fix localities into a particular structural position —accomplices in conservation. The rules of 1931 were based on discussions and experiments in Kumaon between 1909 and 1930 and experiences of collaboration in Burma and Madras. The provincial administration in the United Provinces issued these rules on the basis of recommendations of the Kumaon Forest Grievances Committee with the firm support of the revenue department. Forest department officials came to accept these rules as a reasonable step only after some initial opposition, although the rules did give them an important advisory and supervisory role.

No formal state regulation, however successful, translates into practice exactly as intended. The Forest Council Rules of 1931 underwent the same fate. Different actors involved in forest regulation tried to assert and defend their claims by forcing deviations from intended effects, translating such deviations into accepted norms, and even attempting wholesale changes in the substance of the rules. Over time, claims and counterclaims redefined how the government of forests was to unfold in Kumaon after the 1930s. The most visible aspect of this redefinition was the dispersal of forest regulation to local communities.

The dispersal of regulation did not result in a smooth, seamless mechanism. Surely the creation of many new centers of environmental decision making and delegation of authority has gained the forest and revenue departments important advantages in reducing costs, enhancing compliance, and inspiring the desire to conserve in multiple locations. But it has also produced problems of supervision, coordination, and adequate support for these new centers of decision making. Questions and concerns about measurement, enforcement, supervision, and accountability always accompany delegation. The statements at the beginning of this chapter, which illustrate some of these blemishes, are drawn from interviews with people who occupy a range of official positions. These officers, located at the intersection between what are conventionally taken to be the state and the locality, are a reminder that the concerns of forest conservation have helped undermine and confuse the boundaries between the two entities, if such boundaries were ever precise and clear.

State-Community Relations Prior to 1931

Recall from the discussion in part I of the book that one of the central features of the making of forests was the project of forest improvement to produce greater revenues. The very idea of a forest came to be defined legislatively. The effort to turn land defined as forest into an image of model forests was codified in working plans. Working plans incorporated into departmental regulations the history of the landscape, its prominent climatic and physical features, prevailing vegetation cover, and a plan of action to bring the landscape closer to what a forest should be like. The idea of model forests was implemented even more intensively on plantations through new statistical knowledges about trees, wood, and timber.

All working plans incorporated steps that would limit or eliminate undesired influences in the area of their concern. Many forces affected potential improvements in vegetation. Existing biophysical and edaphic characteristics related to climate, water, and soils were the factors least amenable to tinkering. Human actions were easier to regulate. But ultimately, as Sivaramakrishnan (1999) points out, the cumulative effects of these external influences were impossible fully to envisage, let alone control.

It was easier to exclude and restrict human influences, though this, too, was an immense and complex task. Dangwal, in his original study of the impact of colonial forestry regulations on grazing, argues that "large areas of pastures and sparse population in the precolonial period made it impractical to control grazing" (1997: 411). Attempts by the British to control undesired human influences included such steps as increases in grazing fees, restrictions on seasonal migration of animals, reduction of the number of animals that villagers could graze freely, removal of villagers' cattle sheds, and prevention of the lopping of trees for fodder.[2] In addition, controls over firewood collection, extension of cultivation, the use of fire, and harvesting of timber also became a part of the fabric of administrative control.

As the forest department claimed larger areas, it also grew in size and budget. But its strategy of exclusionary expansion ultimately depended on a balance of threats and sanctions, and it faced opposition from those who viewed these threats and sanctions as impositions. The limits of exclusion were most vividly exposed when villagers acted against the department collectively or where there was widespread individual-level defiance of the new rules. In Bombay Presidency, for example, there were thousands of breaches of new regulations that had reclassified large areas of land as forests (Bombay Forest Commission 1887–88: vol. 4, 37–47, 73–75). The provincial government appointed a commission of inquiry to investigate these offenses as early as 1885. The limits of exclusionary control were equally in stark evidence in Kumaon throughout the early twentieth century.

Prior to the arrival of the British, not only were there few state controls over forests, but village communities also regulated the activities of their residents only sporadically. Neither the rulers of Garhwal nor those of Kumaon extracted much by way of taxation or sale of forest products in the early nineteenth century. There were a few rules to regulate the transhumance practices of villagers when they took their cattle to winter pastures between November and April. According to Traill ([1828] 1992: 66), "this custom has existed from time immemorial, [and] each community has its own particular tract of forest [in the plains] to which it annually returns." He goes on to say that the need to send cattle to the plains did not exist in the northern parts of Kumaon because "the forest lands are more extensive" and there was an "abundance of fine pasture" in the summer months (67). We can conclude that the incidence and intensity of local government of forests was low until the arrival of the British because of the relative abundance of forests, limited uses for many of their products, and poor articulation with markets.[3]

Accounts by British administrators in the late nineteenth century and the early twentieth provide new information about the government of hill forests. In some areas, village-level bodies known as *lattha panchayats* existed.[4] They were usually formed without much state intervention and regulated forest access and use. The more careful land assessments and delineation of village boundaries that the British introduced must have influenced the formation and working of lattha panchayats. Agricultural taxes were collected at the village level and had to be paid in cash. They furthered the careers of certain types of middlemen, among them traders and people appointed as village headmen. The need to make cash payments to the colonial state would likely have prompted greater use of forest products such as green manure to increase the fertility of agricultural fields. As more lands were reclassified and demarcated so as to fall within the administrative boundaries of specific settlements, they also would have been an object of control by villagers.

Over time, the forest department took over larger areas of land and introduced restrictive regulations in other parts of Kumaon. These steps likely boosted self-organization by villagers as well. The scarcities generated by state enclosures of formerly common land had the potential to prompt villagers into institutional innovations to restrict the unbounded use of local resources.[5] S. D. Pant argues that in areas with thick forests there were no restrictions on forest use but in "more populous areas where no such tracts are available, villagers pressed by hard necessity often deliberately let a few patches of arable land lie waste for grazing. A measured plot of land, subscribed by the entire village community is also kept as a grass preserve and constantly watched. . . . This means considerable self-denial and forethought on the part of the village community" (1935: 172). Although these arguments about resource scarcity and the necessity for government are theoretically plausible, the exact relationships between government appropriations of land, re-source scarcities for local populations, and local institutional innovations are difficult to pinpoint in the absence of written evidence.

We do know that several lattha panchayats functioned quite well in the beginning of the twentieth century (Shrivastava 1996: 223–44). Some of them helped protect forests because villagers believed some species of trees, such as deodar, to be sacred (Guha [1989] 2000: 29–30).[6] Long-term economic motivations were at work more often. Many lattha panchayats allocated consumable benefits to villagers and simul-taneously conserved their forests. According to F. Channer, the chief conservator of forests in 1925, community forests near Dwarahat had impressive tree growth in spite of large local demands (cited in Shrivas-tava 1996: 231). Nagarkoti suggests that Chandkot had several well-protected and widely known community forests (1997: 269–70). Some villagers deliberately fallowed land for the production of grasses (Pant 1935: 172). Even after villagers began to create formal forest councils after 1931, there were many lattha panchayats in Kumaon. In the late 1930s, there were nearly twice as many as the approximately two hun-dred official councils (UPFD 1940: 6).

For the most part, the lattha panchayats did not have formal admin-istrative relations with government departments. In some cases, pre-colonial rulers granted lands to villages to be used as commons, but they did not require residents to follow an elaborate set of use restric-tions. Colonial administrators remarked on the fair condition of several forests in which villagers enjoyed the rights of access, use, and manage-ment. But relatively few localities developed an elaborate set of rules to govern forests.[7] By the early twentieth century, residents of one such village, Jalna, near Almora, had developed rules restricting the lopping of oak and cutting of grass and excluding residents from other villages (Pearson 1926: 2–14). In this and other cases, to the extent that local authorities regulated the use of lands around their villages for grazing or firewood collection they did so by creating most of the rules them-selves. According to Pearson (3), "the old customary restrictions on the use of forests had validity and though there was no formal village management, practical protection was largely secured by hill condi-tions and customary limits on user."

Quite apart from their status as informal institutions, the lattha pan-chayats could not work as the basis of a Kumaon-wide network of local regulatory authority because they were too few in number. The costs of knitting them into an effective mechanism of regulation promised to be high given the distances and difficulties of transportation in the hills. But the most important reason why the lattha panchayats were unsuitable as the basis of a new system of finer, more precise, and more intimate forest regulation was because their internal working and customary controls were fluid and highly dependent on the local context. Practices of forest protection embodied in the lattha panchayats can be seen as a species of equilibrium dependent on the needs of villagers, the activities of neighboring villagers in their forests, the rate at which forests regenerated, and the availability of areas to which new demands could be deflected. State-defined objectives about protection of timber species and control of other vegetation played a lesser role. Variations in the resulting makeup, power, and activities of lattha panchayats and changes in their characteristics over time had the potential to vitiate the ability of other villagers to copy what was happening in one location and the ability of state officials to monitor and control. The fact that the existing leadership did not owe its status to state authority also made the lattha panchayats unwieldy instruments of state control. In sum, the diversity of these localized forms of government made them unsuitable for centralized appropriation and guidance.

Indeed, left alone, the forms of social mobilization that the lattha panchayats represented could even act as a foundation for resisting the new regulations that the state introduced. For much of the period prior to the creation of the forest councils, the village community served as a refuge for recalcitrant rural residents. Villagers shielded each other. Village elite did not report on infractions by local residents. Fractious activities within villages caused colonial administrators to complain about the impossibility of apprehending rule violators unless they caught violators in the act (see chapter 6). Existing social organizational forms at the village level thus meant that the village was not a very appropriate intermediary location to facilitate the flow of regulatory authority regarding forests. Rather, the village community acted as an indifferent rival to state power, diverting, blunting, and annulling the effects of state-sponsored control mechanisms.

Colonial administrators did draw some lessons from the lattha panchayats in their efforts to facilitate new forms of government in forests. These lessons were limited to specific activities of the lattha panchayats

and the use of certain types of punitive measures that panchayat decision makers imposed on villagers—social sanctions and exclusion. By
codifying existing local forms of regulations and using the institutional
and political power of the state to leverage their application more
widely, officials hoped to use strategies of government with which villagers were already familiar. Preserving some of the same forms of
sanctions could help to consolidate introduced forms of localized government.

But the greater concern of officials was to reconstitute the lattha
panchayats as formal forest councils joined to the state at the hip. They
accomplished this goal by redefining the relationship between local
government and the state and creating a domain of regulatory authority locally within which coucils would have significant flexibility and
outside of which they would have to depend on formally appointed
state officials. Judiciously managed, the reconfiguration of government
in the localities could allow colonial state officials to bring environment-related interactions within the ambit of state-sponsored resource
control.

Today there are almost no lattha panchayats left. Nagarkoti's recent
study (1997) of eight of these early local forest governments suggests
that where they exist villagers still depend on the forests they manage.
However, few have survived, and most villagers can scarcely remember
that they ever existed.[8]

Discussion

My description of local government of forests in Kumaon prior to the
1930s has emphasized three characteristic features: significant diversity
and limited numbers, incomplete articulation with state authority and
markets, and potential or actual tensions between the interests of state
officials and those of elite and common members of the village community. Before I examine how these aspects of local forest governments
changed, a qualifying note is in order. In focusing on these three features
of protection and regulation of Kumaon forests before their transformation after 1930 and suggesting that the activities of the lattha panchayats
were substantially outside state control prior to the 1930s, even when
they were influenced by official strategies of government, I do not mean
to conform to widely prevalent images that portray Indian villages as

independent centers of civic and political life. Such images have a long pedigree, and they still dominate many studies of rural India.[9]

Consider just one prominent example. Wade's seminal study of community irrigation in South India ([1988] 1994) borrows its title, *Village Republics*, almost in passing from a phrase suggested nearly two centuries ago by Sir Charles Metcalfe of the English East India Company. Impressions of village India as autonomous, unchanging, and self-contained owe their origins to arguments advanced by many colonial observers. Metcalfe discussed "village republics" in a British parliamentary inquiry in 1810.[10] Many other writers on Indian village life relied on his statements, among them James Mill and Karl Marx (Inden 1990: 132; Ludden 1993: 263). For these analyses, Indian villages prior to the arrival of the British were the locus of an autonomous social and political life that flowed independent of macroshifts in rulership and dynasties. Monier-Williams provides a portrait, borrowed almost directly from Metcalfe, writing that the Indian village had

> existed almost unaltered since the description of its organization in Manu's code, two or three centuries before the Christian era. . . . Invader after invader has ravaged the country with fire and sword; internal wars have carried devastation into every corner of the land; tyrannical oppressors have desolated its homesteads; famine has decimated its peasantry, pestilence has depopulated entire districts; floods and earthquakes have changed the face of nature; folly, superstition, and delusion have made havoc of all religion and morality—but the simple, self-contained Indian township has preserved its constitution intact, its customs, precedents, and peculiar institutions unchanged and unchangeable amid all other changes. (1891: 455)

This is powerful stuff. It is also, unfortunately, quite mistaken.[11] It substantiates nicely, in both its assertiveness and its ignorance, Ludden's important point that "colonial knowledge generated authoritative 'facts' that constituted traditional India within a conceptual template that would be progressively theorized within modern world history" (1993: 258; see also Said 1978). Even accounts that question other categories in vogue for depicting Indian politics and history often leave unexamined the idea and formation of "the village."[12] But by now a raft of research has shown the crucial role of macropolitical forces in defining village identity, locating and recording village boundaries, appointing and legitimizing local officials, and influencing positions of informal authority (Ludden 1993; Katten 1999; Stein 1989). Nor did earlier Indian rulers leave peasant society and villages untouched (Stein 1980). It is the nature and forms of state and local interactions that are at stake, not whether they existed prior to colonial rule.

Statements such as those by Metcalfe and Monier-Williams certainly do not hold for Kumaon's villages in general. Where the land revenue administration was concerned, the arrangements initiated by Traill in the 1820s sought to connect village elite squarely with the company's employees. But village elite had enjoyed ties with rulers long before the British defeated the Gorkhas. Traill's measures built on earlier arrangements. In his system of revenue collection, the heads of villages became important intermediaries between the state and the individual cultivator. The presumed antagonism that many historians of village political life have postulated between the state and the village in India thus seems to have little basis in Kumaon in the case of land revenue.

Forests had a different history. A system of state-initiated control and management that relied on detailed plans and strategies was nowhere in evidence in Kumaon before the 1850s. As Baumann remarks, "there is little evidence to support the claim that prior to the British a culture existed wherein 'forest conservation was a social ethic' (Guha [1989] 2000), or indeed of any regular system of forest management in the Uttarakhand" (1995: 63). Many colonial administrators voiced the need to rely on local-level intermediaries. But the reliance on intermediary power brokers did not become official policy before the 1920s. It was only after the 1930s that an organized system of localized decision making for a significant proportion of Kumaon's forests was established. These differences between the government of agricultural and forested land even within Kumaon point to the need for greater care rather than sweeping statements that are supposed to apply to village life generally, even within a region (Scott 1976; Farmer and Bates 1996). It is equally necessary to be circumspect in one's generalizations about political and economic relationships between village communities and state officials.

To conclude this section, the Forest Council Rules and the regulatory effects they produced transformed the landscape of environmental protection in Kumaon. They did so by bringing onto the stage new actors and decision makers—with all their interests, alliances, and practices—and by changing their relationships with existing officials and departments. New rules changed the stakes of village communities and their elite in forests. They also helped shift the balance of villagers' actions in forests from an orientation toward infringing conservationist rules that were an emblem of state policies toward enforcing forest protection rules that came to stand as a feature of local control.

The Forest Council Rules of 1931

The 1878 Indian Forest Act, although it has often been seen as the source of centralized control, also contained provisions for the recognition and formation of village forests (GOI 1890). In Kumaon, the first proposals for constituting community forests were put forward in 1895 by E. K. Pauw (Pearson 1926). V. A. Stowell made another proposal a decade later. He suggested that village communities should have at least some common land that could supply them with grasses and firewood and over which they could exercise proprietary rights. His proposal emphasized the importance of avoiding detailed interference.[13] Shrivastava (1996: 256–57) suggests that this proposal did not travel far because of the obsession of forest department officials with safeguarding state rights in land classified as forests. It required a "detailed count of trees and calculations of outturn on standard working-plan lines" (Pearson 1926), activities impractical at the village level. Later proposals for village forests were often defeated because of fears in the forest department that villagers might assert proprietary rights over the land they gained. To get around such concerns, the commissioner of Kumaon issued a new set of rules in 1924 for creating community forests through grants under land revenue laws (Shrivastava 1996). By this time, many observers had come to believe that the involvement of localities in the government of forests was at least feasible and perhaps even necessary.

Kumaon is not unique in creating some form of community-based government of forests. Cooperatives in Punjab and present-day Himachal Pradesh, village fuelwood reserves in Bombay, forest councils in Madras, and some aspects of the toungya system in Burma relied on the active participation of rural residents in initiatives launched by forest and revenue department officials. Before the Forest Council Rules were prepared in Kumaon, the deputy commissioner studied the system as practiced in Madras and submitted a report on it (GOUP 1927: 2). What does make Kumaon exceptional is the continued survival of localized government of forests, the relatively wide range of villager activities that the forest department was forced to concede, and the autonomy in regulating local actions that the forest councils enjoyed in comparison with similar localized government in other parts of India.[14]

The new rules that initially formed the basis of the changing relationship between localities and the state were a compromise among conflicting interests and claims. They were not to infringe unduly on the subsistence-related activities and harvests of villagers. But the protection of subsistence could not threaten the pursuit of commercial interests by the forest department. The rules had to bind village communities closer to the government and form a wedge between the interests of different actors within the village. They were not intended to turn villages into powerful loci of authority that could compete with the state, but they still needed to lend clout to decision makers within the locality so that residents' activities could effectively be shaped.

As these concerns solidified in written regulations, the need to create new centers of state-sanctioned authority was clear to most colonial administrators. Even the forest department, which was staunchly opposed to handing over forests to district revenue authorities, felt that panchayat forests "promise to be locally of the greatest value to the villagers and generally an important factor in the preservation of the larger forests on which the future prosperity of Kumaun so much depends" (GOUP 1931: 1).

The Forest Council Rules of 1931 contained twenty-two provisions. Of these, eight concerned the formation of new loci of decision making in villages. Nine defined the relationship of the forest councils to the provincial government and the limits within which the councils were to govern forests.[15] The remaining five described the powers forest councils could exercise over their members and their forests.[16] The provincial government modified the Forest Council Rules comprehensively in 1976, aiming to regulate the internal functioning of councils more strictly (Uttar Pradesh Government 1976).[17] By this time, the principle of local, community-level regulation of forests was well established in Kumaon. Table 1 describes and summarizes the main provisions of the Forest Council Rules and their 1976 modifications as these bear on the relationship between the state and the locality.[18]

TABLE 1.
The Forest Council Rules of 1931 and Their 1976 Modifications:
Changes in the Relationship between the State and Communities in Kumaon

Subject	1931	1976
Formation	1. Two or more residents can propose that a council be formed. 2. The council will be formed under the supervision of the deputy commissioner of the district. 3. The boundaries of the forest under the control of the council will be approved by the deputy commissioner.	Rules 2 and 3 remain the same. *Modifications* 1. One-third of the residents must propose the formation of the council.
Membership officials, and decision making	1. All village residents and others possessing rights in the forest will be considered right holders in the council-governed forest. 2. Right holders will elect between three and nine panches as officeholders in the council. 3. The panches will select the council head. 4. The quorum for holding a meeting of council officials is to be two-thirds. 5. All decisions are to be made by a simple majority. 6. Panches can force the removal of one of their members by a majority, and a new member would be elected under the supervision of the deputy commissioner.	Rules 1, 3, 4, and 6 remain the same. *Modifications* 1. Five to nine members will be elected as panches. 2. The deputy commissioner can nominate one member as an official in the council. 3. The head of the council can be removed by one-third of the members, provided their action is approved by a two-thirds majority in the subsequent council meeting. 4. All decisions must be made by a two-thirds majority.
Dissolution	1. The term of the council will be three years, and at the end of that time new elections will be held for officeholders. 2. The deputy commissioner can dissolve a council in cases of repeated mismanagement, and hold fresh elections.	Rule 2 remains the same. *Modifications* 1. The term of the council will be five years.

TABLE 1. (*cont.*)

Subject	1931	1976
Limits on the council's authority	1. Council-governed forest land cannot be sold, mortgaged, or subdivided. 2. Profits from the sale of products from council-governed forests are to be used to improve and safeguard the forest. Any remaining funds are to be used for the benefit of the village community. 3. The council is to demarcate and protect the forest and conserve its trees. 4. The council is to prevent villagers from encroaching on or cultivating in its forest. 5. The council is to maintain records of its meetings and accounts. 6. The council officials are to follow the instructions of higher-level revenue officials.	Rules 1, 2, 4, 5, and 6 remain the same. *Modifications* 1. The council officials must meet at least once every three months. Proceedings of the meeting are to be recorded and a copy submitted to the deputy commissioner. 2. Harvesting of timber beyond one tree each year is to be approved by the deputy commissioner, divisional forest officer, and conservator of forests. Sales of forest products must be in accordance with working plans that the forest department prepares for the council's forest. 3. For commercial sales of forest products, the permission of the divisional forest officer must be obtained. 4. The council must prepare annual budgets and submit copies to the deputy commissioner. These records are to be audited whenever possible.
Supervisory role of state officials	1. All activities related to resin harvesting, except for the domestic use of resin, must be carried out under the supervision of the forest department. Profits from the sale of resin are to be shared in a proportion to be determined by the conservator of forests. 2. Special officials appointed by the provincial government, officials of the forest department, and revenue department officials can inspect the working of the councils to ensure their proper functioning.	Rules 1 and 2 remain the same. *Modifications* 1. Special officers appointed to supervise their functioning must inspect at least one-third of the councils under their jurisdiction each year.

Source: Agrawal 2000.

Three aspects of the rules presented in the table are worth emphasizing. First, the revenue department exercised dominant control over the formation of the forest councils and the term of office of the council's governing committee. Although its control was supposed to be rule bound, the rules gave midlevel officials significant discretionary powers. Second, the authority of the councils was tightly circumscribed where potential conflicts between the villagers' actions and state interests existed. In case revenue and forest department officials did not ensure the election of decision makers who would be sympathetic to the existing aims of governing forests, there were built-in safeguards in the rules that would restrict the extent to which community-level decision makers could stray from the prescribed path. Finally, the forest department retained its commercial interests in the land that came under council authority. That is to say, even if the safeguards failed, the forest department protected its major interests by not decentralizing its commercial rights in Kumaon forests.[19] These three features of the new rules were to ensure that the new decision-making and regulatory authorities in villages would not threaten existing state authority.

The control of the revenue department is most visible in the rules governing the formation and dissolution of the forest councils: these rules show the calculated care that went into granting villagers autonomy but keeping the council officials aware that their selection and status depended on keeping senior officials in the revenue department satisfied. Residents of a settlement initiated the formation of their forest council but only by petitioning the revenue department. Therefore, only villages where there was already some support for community forestry regulations would come under the ambit of the forest council rules. Within these villages, the revenue department could ensure the selection of appropriate officeholders. The government created the new official positions of forest council inspector and officer to facilitate the formation of councils in the 1930s. These officers worked under the supervision of the deputy commissioner. They explained to villagers how the forest councils worked and the advantages of constituting them.

To create their own community forest and forest council, any two residents in a village could submit a petition to the revenue department. The petition launched the processes that transferred land and formal rights to the village. The deputy commissioner ascertained that residents of other nearby villages did not contest the land claims made by the petitioning village. After ensuring that the petition for classifying a certain area of land as community forest was not contested, the deputy commissioner called a meeting of the village residents. Election of a three to nine member governing committee took place during this meeting. Elections were mostly a formality in the initial years. The deputy commissioner selected the members of the governing committee from among the villagers who had assembled. He could dissolve this governing committee on evidence that it was not carrying out its duties well. The committee members could remove officeholders from power, but the selection of new officeholders had to take place under the supervision of the deputy commissioner. The revenue department supervised the working of the forest councils, and the forest department provided technical expertise. The deputy commissioner rendered the final decision on whether and to what extent the area of forests that the villagers desired was to be classified as their community forest. Forests for the councils had to be parceled out from land that was classified as class I forest and was within the boundaries of the village. These three forms of control over the creation of councils and their forests—the constitution of the governing body of the forest councils, the term of its members in office, and the extent of the area that the council was to govern—gave the revenue department a crucial role in the making of forest councils and the territorial extent of their regulatory authority.

A number of rules restricted the powers of community decision makers. Forest councils could not alienate their common land, clear-fell their trees, or appropriate the community good privately. The modifications introduced in 1976 made these intentions of the state clearer. To prevent unauthorized felling of trees, the provincial government introduced debilitating restrictions on the felling of even the reasonable number of trees that the council members might need for domestic use. Villagers had to seek permission from three senior officials to meet their needs for construction timber. Any villager who has had to deal with the Indian bureaucracy knows the sheer folly of imposing such requirements. As one of my informants told me, "Squeezing milk out of stones is not as difficult."[20] Such a rule effectively annulled the ability of villagers to use timber legally. We can safely infer that in the absence of near perfect enforcement all such strict regulations accomplish are rule infractions and side payments.

Two of the rules under "limits on council authority" might seem strangely classified: that the councils should prepare regular records of their meetings and their accounts. However, they indicate the premium that the colonial state placed on increasing the visibility of actions in the village and its forests. Supervision of the forest councils was always sketchy of necessity. The presence of written records facilitated control and supervision over the activities of the councils. It meant that a visiting official could potentially avoid inspecting forests and conducting interviews with villagers by simply inspecting the records, making supervision less time consuming. Of course, the visiting official could always *choose* to examine the condition of the forest or talk to villagers about the functioning of the council. This threat of cross-checking the written word against actual observations ensured that council officials would not simply manufacture written records. The 1976 requirement to prepare budgets and hold regular meetings was an effort in the same direction.[21]

Finally, we come to rules that sought to protect the commercial interests of the forest department even if the first two sets of rules faltered. The forest department had shielded its commercial interests even in the initial classification of lands that were transferred to the revenue department. As it slowly gave up control over the reclassified lands, the forests that found their way to the revenue department and villagers were mainly oak or scattered pine forests, which were difficult to exploit commercially.[22] But even in the forests that councils came to regulate, their actions reduced the costs of enforcement and made commercial exploitation worthwhile. Thus, the Forest Council Rules permitted a refinement of territorial regulation. The forest department (and later the forest corporation set up by the Uttar Pradesh government) continued to be the only agency empowered to harvest resin and timber from council-governed forests. After 1976, forest councils had to involve forest department officials if they wanted to harvest more than one tree a year.

The forest department's hold on commercial benefits derived from the council forests was weakened somewhat by the stipulation that proceeds of timber and resin sales from council-governed forests must be shared with the council, typically to the extent of 40 percent after expenses. The share of the council is deposited with the deputy commissioner and held until the council needs it. The ability of the councils to get these revenues depends on the energy they exercise to obtain the funds from the district coffers. The records maintained by the forest councils indicate that getting the funds out of the deputy commissioner's office typically takes anywhere from two to three years.

The effectiveness of these three types of successive checks on the forest councils and villagers ultimately depends on a range of factors that cannot be controlled by the state. Social and economic differentiation, political tensions, conflicts and the nature of leadership, village size and the economic status of village residents, dependence on forest products, levels of migration and population change, the productivity of given patches of forests, and a host of other factors affect the extent to which villagers are inclined or able to conform to rules. The ability of the forest and revenue departments to monitor compliance across different villages also varies. All of these factors influence whether existing rules will succeed in shaping state-community relations and forest use patterns in intended ways. The flexibility in internal decision making that the rules granted to villagers was no more than a recognition of the fact that the state could not ultimately control all the interactions within the locality.

Effects of the Forest Council Rules

Since 1931, the forested landscape of Kumaon has become densely populated with councils formed under the provisions of the Forest Council Rules, some extremely well managed, others more or less defunct. In the 1930s, the rate of formation of new councils was slow. The appointment of forest panchayat inspectors and officers, on the one hand, increased the intensity of supervision of councils; on the other hand, it also helped spread information about the benefits villagers could gain by forming formal bodies recognized by the government. From less than 400 in 1938, the number of councils increased to about 1,000 in 1955, 1,500 in 1965, and nearly 3,500 today.[23] With just around 12,500 villages in Kumaon, we can safely infer that forest councils exist in nearly a third of them.

A council in every third village—this is a vast network of locally created and enforced regulations compared to the state of affairs under the lattha panchayats or the type of centralized government of forests that the forest department tried to achieve. The lattha panchayats were far fewer in number. The forest department had never reached so far below the surface of village life or shaped so intimately the activities of villagers in forests. Its clumsy efforts to extend the government of forests territorially had backfired, unleashing an orgy of what the Kumaon Forest Grievances Committee was to call "riots and bloodshed" (KFGC 1921).

The involvement of a larger number of Kumaon's residents in the government of forests has produced new splits within the community so that a subgroup has come to occupy the positions of rule makers and enforcers. Whereas the forest department and its officials were visible symbols of restrictions and the target of villagers' ire, now it is no longer possible to consider only the forest department as the source of legal and official constraints on what villagers wish to do in their forests. Villagers, if they wish to protest the restrictions, must do so within the community. But even within communities regulation has many faces. Guards are one of them, perhaps the ones with whom ordinary villagers come into contact most often. Elected officials on the council are another symbol of regulation. But both guards and council officials are also village residents and not always the richest or most powerful ones. The new sources of regulation of forests constitute a different framework for the exercise of power, which produces different effects. It draws attention away from the forest department. It invites change through participation. It dims the possibilities of protest and rebellion. If community-based conservation is active and successful, the prospects of uniting against the forest department become bleak, even remote. Kumaon's colonial forest department succeeded, at least, in blunting the edge of direct protests against its regulatory actions. As Barrington Moore points out in his examination of the sources of agrarian protests, the socially diffused organization of sanctions is nearly immune to the protests invited by centrally administered extractive and sanctioning mechanisms (1966: 459).

The dispersal and increased density of forest government is strikingly obvious. But the real significance of this new government of forests is contained in three additional features: (1) systematization and enhanced certainty, (2) formalization of informal authority, and (3) a certain harmonization of the interests and organization of state and community. These three aspects of the localization of governmentalized forests in Kumaon are likely present in other experiences of localized government as well.

By systematization, I refer to the production of a set of overarching rules that guides the birth of environmental decision making within each village, that streamlines the procedures through which these decision makers are supposed to arrive at plans of action, and that provides to decision-making bodies the same sets of constraints within which they must operate. Decision making, procedures, and constraints existed, of course, prior to the formation of forest councils.[24] But the Forest Council Rules acted as "focal points" (Schelling 1980) for these activities and constraints. Because they had official recognition, conformity to these rules facilitated joint action within the village. Correspondence between these rules and joint action by villagers helped the revenue and forest departments by making it unnecessary to design different strategies to engage diverse local actions. Easier joint action was also in the interests of villagers because it won them assured rights to a specific piece of land that could be counted as their common property. For both parties, uncertainties in interactions and outcomes were thus lowered.

But systematization had another face as well. At the same time that the forest and revenue departments were trying to streamline their interactions with local governments, they were also crafting rules that would permit the local governments significant leeway in their internal decision making. The new rules of government granted villagers flexibility in making decisions that addressed local social, ecological, economic, and political differences. As long as villagers did not contravene the limits within which they were supposed to function, they could craft specific responses to the fluctuations that beset their forests and lives. This dual strategy of standardization in external relationships but diversity in internal activities made for wide variations across councils but limited the problems these variations might produce for the government of forests. There was a finer and more precise delineation of spheres of control and regulation, both territorially and in terms of regulatory authority.[25]

A new formalization of council activities accompanied systematization and standardization. It is visible in the use of prescribed processes, set routines, and predictable methods. Leadership selection through majority rule elections, meetings of elected officials at prescribed intervals, and the official appointment of guards were some such practices. The use of written records is one of the most characteristic aspects of formalization, and forest councils began to produce these records of their activities throughout the length and breadth of Kumaon. Formal records in the same format and for similar kinds of activities across the councils allowed forest council inspectors and officers to keep up with developments in their forest councils with a minimum of energy and time. Greater formalization, especially through the production of written documents, permitted diversity to flourish, but it also tamed the effects of such diversity by letting government officials understand it more easily.

Formalization also characterizes the new relationship between state officials and community-level decision makers. Prior to the formation of the forest councils, village communities interacted with state officials mostly on an informal basis. There were no set channels for the flow of regulatory power. The creation of the councils made state-locality relationships far more formal. At the same time, new procedures and routines help generate asymmetric flows of power between appointed state and elected council officials. Council officials typically approach state officials when they need something: recovery of fines and dues, seeds or saplings for afforestation, permission to cut a tree, and so forth. By creating a one-to-one relationship between each forest council and state officials, and by making no provision for interactions or alliances among the forest councils, the forest council rules have consolidated the political inequalities between state and locality.

Although individual localities and the state are fixed in a structurally unequal relationship, the antagonisms between them have been radically reconfigured through decentralized conservation. Under the new system of governing forests, state power is more deeply entrenched and localized authority is stronger. The two reinforce rather than oppose each other. State support of local decision makers in the form of advice, the apprehension of rule breakers, the collection of fines, the planting of new vegetation, and manifold other activities strengthens processes of enforcement and forest conservation—the common interests of both local and state officials. For state officials to weaken community-level decision makers by withholding such support only detracts from the objective of environmental protection.[26] Protests against state officials organized by local leaders only weaken the ability of both to enforce regulations.

The formation of the forest councils extended the government of forests that the creation of provincial forest departments had initiated in the previous century. The forest departments facilitated what might be called "government at a distance" (Latour 1987; Miller and Rose 1990) by adopting formal statutes and working plans to guide their activities and employing statistical measures and monitoring mechanisms to evaluate the success of their plans and to recalibrate them. The ability to transform the world of vegetation into numbers and the relationships between different features of the landscape into equations meant that superior officials could create numerical measures of performance and evaluation. By giving subordinates numerical performance measures, the forest department tried to reduce ambiguities in the tasks to be carried out and targets to be met. Efforts to exercise power from a distance and shape the actions of subordinates succeeded in part because of the acceptance of the goals of forestry as appropriate and in part because the interests of subordinates were linked to realization of their targets.

But these clear targets and unambiguous measures failed to accomplish the desired conversion of the landscape in the face of determined resistance by villagers. Recourse to brute enforcement proved ineffective. Villagers did not see why they had to change their behavior in forests to conform to the objectives formulated by the forest department. The costs of monitoring and sanctioning villagers were high enough, but even more costly were the potential political effects of strict enforcement. Villagers deprived of the basic means of their livelihood would likely come to distrust the entire colonial administration. Successful "government at a distance" required the restructuring of interests in forest conservation. Only when villagers saw forests as theirs and the condition of forests as dependent on their actions would they begin to follow protectionist strategies. The formation of forest councils and the handing over of forests to them were steps seeking to accomplish such a restructuring of villagers' interests. In the process, they also established complementary but asymmetric flows of support and dependence as the chief characteristics of the political relationship between the forest councils and provincial government officials.

Conclusion

This chapter has examined the first of the three sets of changes in relationships and subjectivities that comprise the decentralization of environmental policies. The reconfiguration of interests in forests that accompanied the transformations described in this chapter was a result of two significant developments. The first were the concrete struggles that villagers waged in defense of continued rights over forests. The second was the dawning realization among provincial government officials that centralized control over forests would prove prohibitively expensive to maintain. New technologies of government that bound localities to the defense of forests turned out to be far less costly, both politically and economically. It is this recourse to localized government that resulted in greater predictability of environmental decision making through systematization and formalization.

One way to think about the changing environmental relationships between state officials and villages in Kumaon would be to call them state formation (Agrawal 2001a). State formation in this context corresponds to activities that contribute to the formalization and systematization of social action and in so doing consolidate or complicate the division between states and societies.[27] Examples of such activities in Kumaon are the creation of new rules to define the limits of the permissible; the institution of new organizational structures to enforce such rules; and the incorporation, and thereby undermining, of alternative loci for the exercise of power. State officials, one might argue, increasingly become the interpreters and enforcers of what is permissible as state making proceeds apace.[28]

As a descriptor of what happened in Kumaon, however, the idea of state formation needs further work. Two basic problems make the phrase inappropriate: (1) the idea that something like a state is being created and consolidated, together with the usual assumptions about states such as their monopoly of control over the means of violence; and (2) the notions that states and localities are opposed to each other and that the power of the state is aimed at the control of localities. The second assumption often involves the corollary suggestion that state formation involves a displacement of existing local relations and forms. Ferguson's study of development in Lesotho thus suggests that development is "not a machine for eliminating poverty that is [also] incidentally involved with the state bureaucracy; it is a machine for reinforcing and expanding the exercise of bureaucratic state power, a state strategy whose principal effect is expanding and entrenching state power" ([1990] 1994: 255). He qualifies his arguments by drawing on Foucault, suggesting that the state is not a unitary actor and that the expansion of state power implies that "specific bureaucratic knots of power are implanted [into society? community?], an infestation of petty bureaucrats wielding petty powers" (273).[29] But the idea of state formation and expansion ultimately underpins his critique of development.

The Kumaon case is different. For one thing, the Forest Council Rules did not make localities into bureaucratic mini-images of the state. New technologies of environmental government, pursued by the local decision makers and state officials jointly, led to a greater systematization of rules and the production of written records, but they did not mean that localities turned into formal, impersonal social spaces. Rather, strategies of government relied on existing forms of cooperation and joint action. New processes of government came to exist in conjunction with other social processes within the locality and often drew strength from them. Indeed, the complementarity between the dispersal of protectionist strategies to the village level and the existing strengths of local governments is precisely the reason why forest councils were born. As the previous section tried to establish, the strength of localities was linked to that of state officials in a positive manner. In the next chapter, we shall see how villagers' actions in forests changed and the differences in the responses and burdens on villagers that were a part of new forms of government of forests.

Instead of viewing forest councils as part of a process of state formation, it is perhaps more appropriate to see them as a form of government that encourages (and depends for its success on) the willing participation of those subject to rule and rules. Although strategies to force compliance are present in the repertoire of governmental mechanisms that forest councils can use, their deployment is rare at best. The dominant idioms employed to secure compliance within the community are those of cooperation, achievement of group interests, and safeguarding the future. In contrast, one might point out, almost all the definitions of the state typically draw attention to means of coercion and monopolies over violence. Government, on the other hand, shifts attention to the multiple other means of shaping behavior that are at the disposal of power.

5

Inside the Regulatory Community

> Virtue regenerated—crime reduced—public safety enhanced—institu-
> tionalization banished—dependency transformed to activity . . . politi-
> cal alienation reduced . . . the Gordian knot of state versus individual not
> cut but untied, all by a simple idea in politics: community.
> —Nikolas Rose, *Powers of Freedom*

By the 1940s, community-based forest protection was firmly in place in
Kumaon, even though the number of councils was relatively small. As
government officials and village leaders came to accept the permanence
of this new form of regulation over the next few decades, thousands of
new councils were born. They dispersed government throughout Kumaon. Decision makers in these community-based councils became
agents of environmental regulation. Many of the processes that the
forest department had tried to impose on the social body of Kumaon
and through which it had tried to shape ecological outcomes became
part of the regulatory charge of communities. The joint government of
forests turned out to be more intimate, far reaching, penetrating, and
effective than anticipated by the critics of community regulation and
even by many of its sympathizers.

The new forms of regulation that community-based councils used in
defense of forests were as diverse as the means villagers found to evade
regulation. Some forms of environmental regulation were invented
afresh in response to the new powers community decision makers
gained. But new regulations of villagers' activities in forests were often
drawn from existing practices used informally to curb deviance and
promote compliance in other spheres of social interaction.[1] For example, informal village councils chastised community members in front of
the entire settlement if they violated existing social norms. Such punishments could easily be deployed to deter illegitimate actions in the forest.
Prevailing forms of interaction could also assume new valence in the
context of environmental regulation. The fact that villagers could see
each other and had intimate knowledge of the activities of other house-

holds made monitoring far less costly than the strategies pursued by centralized government.

Greater diversity and economy were only two of the characteristics of community-based allocation, monitoring, and sanctioning. There were other aspects to local regulation. It was more autonomous and legitimate. It was also more continuous and modulated. Its autonomy and legitimacy derived at least in part from its closeness to the objects of regulation. Villagers themselves selected (or rejected) those who would regulate. The actions and decisions of regulators were less hidden from them than had been the case when forest department officials were exercising control. The continuity and modulation of local regulation stemmed from the better understanding about their locality that the new regulators possessed. After all, officials of the forest councils and those they sought to regulate both lived within the bosom of the same community.

The transformations of social relations within the community, the birth of new forms of environmental regulation, and the juxtaposition of new characteristics of these forms of regulation in the community can be seen to mark the constitution of a new instrument for the production of environmental conformity: the regulatory community. The reconfiguration of the localized community in its regulatory form occurred with the help of colonial state officials as a search for a new way to govern forests. The reconfiguration continues to unfold today. The ensuing discussion examines mainly the political and economic effects of regulation. However, environmental government also has a productive and positive effect on forms of subjectivity. Regulations act as the crucible in which environmental subjectivities are forged (Moore 1999).

The Crucial Role of Local Regulation in the Government of the Environment

The Hobbesian state of nature has often been taken to be a reasonable approximation of the obstacles facing those who seek to govern valuable environmental resources such as forests.[2] Under such conditions, proponents of Malthus would argue for the impossibility of successful, sustainable government of nature. Present-day Kumaon contains all the ingredients that would have made Malthus throw up his hands in despair: increasing population pressures, a dense network of roads, significant articulation with market demand, and high levels of dependence on forests among local residents. In the face of these diverse pressures to consume fodder, firewood, and timber and nontimber forest products, the need for regulatory regimes to restrict harvesting and use may seem unexceptional. Indeed, the intensity of these pressures even raises questions about the extent to which localized, institutionalized regulations can be effective in efforts to protect and conserve. On the resolution of doubts about the efficacy of local regulation hinge the prospects of a vast number of national initiatives to protect resources through the community.[3]

Several competing explanations attempt to account for variations in local resource conditions. Among the most frequently cited factors are market demands and population pressures. These two forces have long enjoyed a favored status among those interested in understanding passages from antiquity. Marxist understandings of modernization, structuralist writings on agrarian transformations, and institutionalist analyses of economic shifts have all looked to population increases and market integration as a motor of change. Even technological innovation, that engine of growth, is often traced to market competition and population growth. Necessity may be the mother of invention, but it may itself be engendered by population increases and greater market articulation.

But where resource use and regulation are concerned the role of population is typically seen as negative. Higher levels of population are supposed to contribute to deforestation, soil degradation, reductions in biodiversity, food scarcities, and global climate change. An immense and impressive scholarship thus explains how increasing population pressures have contributed to environmental degradation.[4] Such conceptualizations of the links between population and environment, as Arizpe and Valasquez (1994: 1) point out, pose a stark choice between people's needs and conservation of the environment. They relate population pressure to environmental degradation in a rather straightforward fashion.[5] In so doing, they ignore the complexities that politics, regulations, and social interactions introduce in the use and government of resources, as has been pointed out by a number of scholars (Geist and Lambin 2001; Ives and Pitt 1988; Tiffen, Mortimore, and Gichuki 1994).

Assessments of the relationship between market pressures and environmental change are also usually negative. In the context of Kumaon's forest councils, the perspective of neo-Smithian Marxists, to use Brenner's (1977: 26–27) evocative phrase, holds obvious implications. As local economies are integrated into larger markets, greater market pressures are supposed to create higher rates of deforestation. Local users are more likely to increase their harvesting levels since they can now exploit resources for cash as well.[6] The role played by roads and better transportation links is viewed as critical in this regard.[7]

Many of these arguments about population and markets draw their inspiration from macrostructural relationships. They do not satisfactorily explain microlevel variations in how people protect, govern, and consume environmental resources. The question is whether different users have the same response to given changes in demand and other pressures irrespective of their contexts, institutions, histories, or beliefs and whether increasing resource pressures reshape even institutions and beliefs in a convergent fashion. One can reasonably claim that resource-related outcomes of all human actions are mediated by regulatory arrangements that soften, attenuate, structure, mold, or accentuate results.[8] Under their influence, users may forgo cash incomes, divert their demands, reduce levels of exploitation, or begin to rely on substitutes. Community-level decision makers in Kumaon not only create and enforce regulations but also demand support from government actors in the forest and revenue departments to deal with those who do not comply. Nor is Kumaon exceptional in its regulatory rules, as the immense literature on community-based conservation makes clear.[9] Those who believe in community argue that communities and similar small social formations can create and sustain institutionalized patterns of interaction to govern collectively owned resources successfully, even in the face of adverse pressures from states, demographic changes, and market forces.[10]

The forest councils in Kumaon form an ideal empirical case study through which to examine the extent to which localized regulation and variations in the nature of regulatory intensity affect resources, especially in the face of changes in population pressure and new articulations with markets. The councils are located in a broadly similar cultural and technological environment. They face the same set of state institutions and actors—the Forest Council Rules, the forest department, and the revenue department. But they differ in levels of local regulation and also in their size, population, social stratification, and distance from markets and the degree to which their interactions with state officials have been formalized.

I collected information for 279 forest councils on such factors as population pressures, market forces, age of councils, frequency of local meetings, and the length of time each year that the councils guarded their forests. This information formed the foundation of my statistical analysis of the relative impact of regulation on the condition of forests. The results show the extremely important role of local regulation in the government of forests and in variations in the condition of forests across the councils.[11] On the average, councils and villagers held nearly seven meetings per year to discuss their forests and used different forms of monitoring and guarding their forests for more than half the year. Of the factors mentioned above, per capita land availability, the age of a forest council, the subsistence value of the forest for villagers, and the amount of time during which councils guarded the forest emerged as having a positive relationship with the condition of council forests. As the value of these factors increased, the likelihood that villagers' would assess their forests as being in a good condition also improved. But the most important factor in affecting the condition of the forests turned out to be the length of time during which villagers guarded forests. The magnitude of the effect of this aspect of regulation was larger and statistically more significant than the effects of any of the other factors.

Indeed, this finding about the importance of regulation in community-level government of forests is not surprising. Many studies of local use and control over environmental resources have reported on the need for robust regulation to ensure continued survival of resources. Wade's work on village-level irrigation institutions in South India ([1988] 1994), McKean's investigation of forest use in medieval Japan (1992), Acheson's study of lobster fisheries in North America (2003), Ostrom's examination of a range of resources around the world (1990), and Baland and Platteau's review of the case literature on commons (1996) all focus on regulation as a critical element for successful local government. These and other studies also underline the importance of appropriate allocation, information collection and monitoring, and sanctions in making local government of resources effective.

Regulatory Strategies

Historians and anthropologists have long examined localized regulation of socioeconomic and political dynamics in the context of land and resource use. But the regulatory regimes on which they focus have typically seemed antiquated, exotic, or both. Portrayals of the most famous example of community control over the use of land—the English Commons (Ault 1952; Thirsk 1966; Yelling 1977)—suggest, often only by implication, that common property is a curious holdover from the past that was destined to disappear in the face of trends toward modernization. The processes of social-ecological regulation that the Kumaon forest councils embody, however, seem thoroughly modern. They emerged as a result of intimate interactions between state officials and local residents.

The critical role of community regulation in the government of forests in Kumaon makes it important to examine *how* it works and with what effects. The new regulatory charge of communities is not just a negative force: preventing villagers from doing some things, constraining them to certain types of actions, and in general restricting their freedom in the forest. It does some of that, of course, but it is simultaneously enormously productive.

The dispersal of regulation in the 1930s gave birth to a whole range of new forms of allocation, monitoring, sanctions, and adjudication. Earlier, the forest department had at its service the powerful tools of science, hierarchy, and law. But ultimately the instruments it used—statistics and numbers, rules and ordinances, fines and imprisonment—were blunt in comparison with the tremendous ecological and social diversity they sought to tame. Regulatory communities operate at a finer level of social interaction. They deploy a more precise set of instruments, which can be elaborated and calibrated with much greater detail. These instruments help make visible much that is hidden from the official eye of the state and transform residents' views about ecological practice.

The deployment of these new strategies of monitoring, sanctioning, and adjudication created additional conflicts in political relations within the community. When villagers constituted themselves as an organized body to protect forests with the help of state officials, they simultaneously created one social group comprising rule makers and enforcers and another that included rule followers and breakers. By setting one group of villagers against another, the creation of community-based regulation eased the enforcement burden of the forest department. In the new alliances that came into being, villagers could neither rely on their leaders to struggle against state-appointed officials nor contest control over forests. Their leaders had become representatives of ecological enforcement. Many villagers themselves began to argue for the importance of regulation and enforcement in favor of protection of forests. As the burden of regulation fell unequally on different groups within villages, they also began jockeying for the new positions of power and enforcement.

As pursued by the forest department, the regulation of human actions in forests had two major components: invasive and restrictive. By classifying selected areas into circles and compartments, department officials readied them for specific invasive regimes of treatment and transformation. Planting, thinning, clearing, and improvement-felling operations shaped the existing vegetation toward the desired end of higher and regular yields of wood volume. In contrast to these invasive interventions, the forest department sought to regulate human influences through restrictive measures. It attempted to limit grazing, firing, lopping, and fodder and firewood collection.[12] The department found it difficult enough to deal with the volatility of climate, the variability of soils, the virulence of insects and pests, and the consequent unpredictability of vegetation growth. But its restrictive efforts to tame human actions only produced exceptional deviance and unparalleled illegality.

The councils faced the same dilemma as did the forest department but with a twist: how to achieve effective regulation of actions that according to some department officials would wipe out all forests if unchecked and do so without eroding existing relationships within the community. Further, regulation had to be accomplished in the aftermath of the failure of the forest department and without alienating its targets. Since the target and source of regulation—villagers and the forest councils—were connected through a whole network of socioeconomic and other relationships, strategies that would alienate targets of regulation from their neighbors could turn out to be extremely costly socially.

The important elements of regulation in Kumaon—allocation, monitoring, and sanctioning—found practical expression in multiple forms. In each case, decision makers in communities took the provisions that the Forest Council Rules of 1931 specified and elaborated and adapted them to the particular social and ecological contexts of their forests.[13]

The set of codified rules that the central government created (see table 2) turned out only to be a basis for further experimentation. State officials could not anticipate all the circumstances of use and management at the local level. Their silence, because of the human inability to anticipate perfectly, meant that the written rules worked in conjunction with local practices instead of marking the end of local institutional innovation or an absolute limit on the evolution of existing practices. They created a space for regulatory activity that the forest councils filled with their interpretations and decisions. Indeed, as Wittgenstein has demonstrated, all rules, however detailed, require interpretation in light of background understandings and practices that can never be embodied within the rules themselves.[14] In Kumaon, villagers' needs, social stratification, power relations in the community, levels of migration, and a whole host of other factors affected how councils could shape villagers' actions in forests.

Allocation

The entries in table 2 under "Allocation" show that the provincial government made few specific rules about how the forest councils were to allocate forest products. Forest councils gained the power to assess the condition of their forests and allocate forest products among their right holder members.[15] Even the modifications of 1976 sought mainly to skim a share of commercial revenues from forests. They did not significantly alter the contractual nature of regulatory authority within the community and between community-level decision makers and village residents.

The significant discretion granted to forest councils in the way they allocate resources has resulted in wide diversity in patterns of harvests and appropriation. Villagers have used the forest councils to create

TABLE 2.

The Forest Council Rules of Kumaon, 1931 and 1976:
Relations between Communities and Their Residents

Subject	1931	1976
Allocation	1. Councils can regulate grazing and the collection of fodder, nontimber forest products, firewood, and construction stones in forests. 2. Councils have the power to distribute forest products to right holders. 3. Councils have the power to distribute and sell forest products to those who do not hold rights.	Rules 2 and 3 remain the same. *Modifications* 1. At least 20 percent of the forest should be excluded from grazing each year. 2. Land may be leased for commercial purposes. 3. No more than one tree may be granted to a right holder without the consent of more than half the members of the forest council's executive committee.
Monitoring	1. The council can appoint guards to monitor and enforce rules.	*Modifications* 1. Appointments of guards require approval from the deputy commissioner.
Sanctioning	1. The council can impose fines on rule violators up to Rs. 5. 2. Fines imposed by the council are to be treated as government dues and are recoverable using similar procedures. 3. The council can impound animals found grazing without its permission. 4. The council can confiscate cutting implements from right holders present in the forest illegally. 5. The council can restrict or suspend the rights of those who break rules regularly.	Rules 2, 3, 4, and 5 remain the same. *Modifications* 1. The council can impose fines up to Rs. 50 with the agreement of rule violators, and up to Rs. 500 with the permission of the deputy commissioner.

TABLE 2. (*cont.*)

Subject	1931	1976
Adjudication	1. The council can decide on the level of punishment of persons found to be breaking rules. 2. The council can file court cases in situations in which it considers higher levels of fines necessary.	Rule 2 remains the same. *Modifications* 1. Court cases can be filed against habitual rule violators.
Income	1. All income from the sale of noncommercial forest products is to be assigned to the council. 2. Income from the sale of resin and timber is to be allocated to the council in a proportion determined by the conservator of forests (in practice it has been allocated fully to the councils). 3. The council can allocate income to right holders.	Rules 1 and 3 remain the same. *Modifications* 1. The forest department can deduct 10 percent of the gross commercial revenues of the councils to meet expenses. 2. Net income from the sale of timber and resin is to be deposited in a fund managed by the deputy commissioner. 3. 20 percent of the net income of the district council is to be allocated to development. 4. 40 percent of net income is to be allocated to the forest department 5. 40 percent of net income is to be managed by the deputy commissioner and allocated to councils for local projects.

Source: Agrawal 2000.

different institutional mechanisms that define who can take what from where at what time and for what purpose. The "who" refers to "right holders in the forest" (Sarin 2002). They may be defined by gender, residency status, wealth, livestock ownership, or contributions to forming the forest council. The "what" includes fodder, grazing of livestock, firewood, timber, medicinal plants, mushrooms, edible berries and fruits, and construction stones. Allocation regimes vary by product. Some products can be harvested year round or whenever they are in season. Others are limited by size, amount, volume, types of harvesting tools, and species.

Diversity also characterizes the referents of "where" and "when" and the uses to which harvested products are put. For example, many councils divide their forests into several compartments. Villagers can harvest fodder or graze animals in a given compartment only in specified years. But because of limited regeneration the amount of time in which villagers are free to extract fodder varies anywhere from two weeks to four months. Villagers can also cut leaves and lop small branches as fodder from certain species. In some cases, councils frame rules that prescribe whether right holders can sell their shares. A large number of councils do not permit their right holders to use firewood in tea shops or for smithing. Most villagers use products from council-governed forests domestically.

Despite some broad similarities and constraints, there is tremendous variation in allocation rules. Different rules along various dimensions can be combined to yield literally hundreds of feasible combinations.[16] Yet we must acknowledge that the creation of forest councils has led to the streamlining, regulation, and standardization of an even greater diversity of use patterns and everyday practices that must have existed already. Consider spatial limits as an example. One significant underpinning of all council actions is the territorialization of harvests from the forest. Defining clear boundaries in the forests belonging to a given village means that village residents can legally harvest products only in fixed areas. A second form of reduction in diversity concerns how the council's executive committee documents its own decisions, villager actions, and the state of the forest. The legalization of councils' decision making goes hand in hand with formal documentation, which makes rule design, enforcement, and infractions visible to powerful outsiders. A third example concerns the extent to which forest councils can market products such as timber and resin. Now they must involve the forest department.

Different forms of allocation result in varying ecological and distributive consequences. To gain a better sense of such differences, take the case of fodder. The example of this specific commodity clarifies some of the differences between community and forest department regimes of allocation. Fodder, whether grass or leaf, is a renewable resource. To ensure regular annual supplies, it is necessary to match extraction and regeneration. Community officials in well-functioning forest councils do so by assessing fodder growth during the year, fixing extraction quotas below estimated levels of annual regeneration, and monitoring fodder extraction using simple, easy to understand methods.

Under one procedure of allocation that is followed in a number of villages, fodder is distributed equally among all right-holding households. Council officials make an eyeball estimate of available fodder. The total number of animals that can graze or bundles of fodder that can be harvested depends on this initial estimate. The figure is divided among village households equally, and each household receives a pass specifying the exact number of bundles or animals to which it is entitled. Forest guards or council officials monitor extractions. Bundles of grass are measured with the help of a uniform length of rope, which each harvesting household is expected to use after cutting its share of grass. The number of animals is counted. Villagers cut fodder or graze their animals on days specified by the council. On these days, council officials are present in the forest compartment to ensure that no one harvests an amount greater than his or her share. A new compartment is opened when the grass in the previous one is exhausted. These steps enhance uniformity and equality in allocation.

A second procedure in which the councils attempt to match withdrawal to regeneration involves the allocation of patches of forest floor to households. Villagers assemble on a specified day near the compartment that has been opened for harvesting. At the appointed time, they begin cutting grass from their assigned patches. Once they complete the harvesting, they are free to take their bundles home. This method is used more commonly for harvesting fodder than for grazing animals. It rewards households that can cut large quantities of fodder in a given time span.

Under a third scenario, grass or tree fodder in the community's forest is sold at an auction. This means of allocation does not lead to a close match between withdrawal and regeneration. The auction winner is free to harvest as much fodder as possible from the compartments for which he has made a winning bid. Cutting grass too close to the ground, grazing animals for too long a period, or lopping branches that are too large adversely affects future regeneration but allows the winning bidder to gain a larger allocation in the current year. This third allocation procedure has highly unequal distributive consequences.

Among the thirty-eight Kumaon villages where I conducted research between 1989 and 1993, one or another of these three forms of fodder allocation was commonly used. There were variations in timing, type of auction, openness of auctions, and the role of council officials, but the three patterns held broadly. The great majority of forest councils (twenty-six) allocated fodder equally. Allocations of fodder by space was the least common (three councils). The remaining nine councils auctioned fodder. Councils often experimented with different procedures before settling on a preference.

The Bhagartola and Majethi councils in Almora district are good examples of the two main forms of allocation: auctions versus equal numbers of bundles. The two councils are similar in size (in population and forest area) and distance from roads. But they differ in the composition of the households and the amounts they spend on monitoring. Villagers formed the Bhagartola council in 1937. The settlement contains households belonging to three different castes: *brahmans*, *thakurs*, (upper castes) and *harijans* (lower castes). There are no obvious group conflicts in the village. The council spent 3,100 rupees a year (approximately U.S.$150.00 at prevailing rates of exchange) on the average on guard salary between 1977 and 1992. The second village, Majethi, comprises mainly brahman and harijan households. The lower castes are outnumbered, and there is a long history of simmering hostility between the two groups. The five-member executive body of the Majethi forest council was formed on the initiative of the brahman leaders (1961). It has only brahman members and spends little on a guard.

The Bhagartola council allocates fodder equally among its members. Its forests are divided into three sections, which are opened to fodder harvests in turn. Animals are not allowed to graze in the forest, but over the course of the year the council permits villagers to enter the forest and harvest bundles of fodder for six to twelve weeks. Villagers have equal rights to fodder bundles. Each household sends one person to cut fodder. Villagers congregate near the forest in the morning. After a council official declares the forest open to harvesting, each villager harvests the allocated bundles. In Majethi, fodder from the forest is auctioned to the highest bidder. The forest is divided into four sections, and the grass from each section is auctioned separately. Between 1961 and 1991, brahmans bid successfully for the right to harvest fodder from every section of the council forest. The harijans in the village, if they wanted fodder, had to buy it from the individual who had won the auction. Harijans in the village have recently tried to form coalitions to bid in the auctions, but none has yet bid successfully.

Auctions as a form of allocation are associated with unequal distribution and low enforcement costs. Once a council has auctioned its fodder, it need no longer worry about regulating illegal extraction. The winning bidder ensures that no one else cuts fodder or grazes animals in his section. Auctions also concentrate benefits. The villages that have adopted auctions as a form of allocation are typically polarized between different social groups, and tensions between groups are frequent.

The diversity of allocation regimes at the community level distinguishes them from the forms of allocation that the forest department deployed. For one thing, the department was interested mainly in products that yielded revenue. Timber was its main preoccupation, and it attempted to minimize harvesting costs, including those of monitoring and supervision. Villagers were typically not permitted to harvest timber in certain forests or of certain species. For grazing, they were charged a flat fee. In contrast, community-based allocation concerns a greater range of products and produces a larger variety of rules for each product. The first corollary of this diversity of rules is informational. Members of the community know the many different rules and expectations that govern their activities. The second corollary is related to monitoring. The enforcement of elaborate procedures and detailed rules is dependent on the existence of equally sophisticated monitoring mechanisms. The third corollary concerns the economic implications of applying diverse rules to resources that do not have a high commercial value. The cost of monitoring and enforcement should not be high.

Centralized enforcement of proliferating and increasingly detailed rules, however, produces perverse effects. Either centralized supervisory and monitoring mechanisms are overwhelmed when they are required to demonstrate such fine discernment for low-value goods (because returns from more precise monitoring are too low compared to costs) or, more likely, they become vulnerable to corruption. Those with the power to enforce use it to extract side payments. The political relationship between those who monitor and those who are monitored needs then to be equitably reciprocal, not highly asymmetrical.

Monitoring

The ability of the forest department to control its vast estate was translated into practice by the guards it hired. The effectiveness of its scrutiny was limited by the vast areas it controlled and problems in ensuring that the appointed guards followed the letter and the spirit of the allocation mechanisms it wanted to implement. In locations where there were no conflicts with existing users, the department only had to be concerned about whether its personnel were carrying out the assigned tasks of surveying the land and vegetation and undertaking the required planting, thinning, felling, and transportation tasks. But where departmental operations were in tension with existing use patterns the complexity of the monitoring task multiplied.

Recall from the discussion in the previous chapters the high numbers of legal infractions that departmental staff detected in the forest. These cases concerned fires in the forest, illegal grazing, and removal of firewood, fodder, and timber without permission. For India, the total number of detected violations grew from around 5,000 cases a year at the beginning of the twentieth century to nearly 150,000 a year by the early 1940s. The rate of growth of violations was lower in Kumaon; their number increased from under 1,000 in the late nineteenth century to about 2,500 by 1920 and remained at that level until the 1940s.

The department's options for addressing the increasing number and complexity of infractions were limited: it could increase the number of guards and/or increase their remuneration. The first option enhanced the supervisory burden. Who would monitor the monitor? The effectiveness of the second option—higher salaries—was also doubtful. In addition, it had a direct impact on the profits of the forest department. Ultimately, neither of the two options proved practical.

In common with the forest department, the forest councils face many problems in securing the compliance of ordinary village residents. The councils draft the basic rules to govern the use of their forests within a year or two of formation and election. After that, their key activity is monitoring and, where possible, limiting violations of protective arrangements. But violations of the rules they create are common. In the villages in which I conducted research, villagers often illegally entered the council-managed forests, cut grass and leaf fodder from trees,

TABLE 3.

Detection of Rule Infractions by Forest Council Guards
in Kumaon, 1977–92

Village	Number of Meetings Held per Year	Average Annual Amount Spent on Protection (Rs.)	Average Annual Number of Detected Rule Infractions
Airadi	3	790	32
Banua	6	2,835	156
Bhagartola	10	3,100	192
Ladfoda	3	2,840	121
Miraini	8	832	62
Lohathal	4	1,850	109
Nagilagaon	5	2,500	89
Tangnua	4	175	8

Source: Field research, 1993.

Note: The figures are for seven years selected at random from the records maintained by the forest councils.

grazed their animals, gathered green manure from the forest floor, picked up twigs and fallen branches, gathered construction stones, and, rarely, cut a tree or two. These activities were not permitted and occurred despite the knowledge of villagers that if detected they would face a variety of sanctions.

Table 3 presents information on infractions committed by residents of eight villages. The data are from written records maintained by the forest councils at the village level. The information in the table can be used to make several points, two of which deserve initial attention. First, the level of infractions within communities seems to be extremely high. Second, there is a close relationship between the efforts councils devote to monitoring and the level of detected infractions listed in council records.

The records show that many villagers violate existing arrangements for the allocation of forest products and often commit violations. Nonetheless, the written records almost certainly underestimate the extent of illegal grazing and cutting. To detect all violations, all behavior in the village must be monitored—a prohibitively expensive proposition. Persons who are supposed to monitor the forest are often absent, careless, or unable to cover all the compartments of the forest at the same time.

The information in table 3 comes from a subset of the councils that maintained records from among the thirty-eight I studied. Several councils did not maintain careful records or hold regular meetings. But where records were kept and where councils tried to enforce their rules, rule breaking seemed to be endemic. The table suggests that in such cases the average number of detected violations was nearly ninety per year. The councils use many methods to monitor villagers' actions in the forest (see below), and the costs of the different mechanisms they deploy vary significantly. But whatever the mechanism it is a safe assumption that many actions in the forest remain hidden from the official view even at the level of decision makers within the community. The tea shop owner in Bhagartola, who relies on firewood from the forest to keep his stove going, said as much when he described the way the world works: "To what extent can you keep watch over the forest? Nowadays even one's own property is not safe if your eyes are not on it twenty-four hours. The forest is big, and there is just one guard. How far can he make rounds of the forest?"[17]

Although there is no way to know for sure, conversations with the villagers left a general impression that even well-functioning councils detect only a fraction of all rule-violating behavior. I interviewed more than two hundred villagers in thirty-eight councils. With remarkable regularity, they asserted that the council (and its guards) had been too hard on them, was not aware of offenses committed by their neighbors and friends, and was too lax in controlling firewood and fodder theft by other villagers! Villagers who had not recently been caught breaking council rules also pointed the finger at numerous families whose rule infractions had gone undetected. These conversations suggest that councils and their officials uncover perhaps a quarter to a fifth of all violations. The eight villages in table 3 vary in size, the amount of forest they manage, their proximity to markets, and their levels of out-migration. As a group, they are not visibly different from other villages in the hills. If the figures for these eight villages resemble what happens in villages in Kumaon in general, the total number of infractions per year in Kumaon seems strikingly high. For the twelve thousand or so villages in the three districts of Kumaon, even the detected rule violations add up to more than a million instances of rule breaking each year. And this may be only a quarter of all rule violations.

Compare this figure with the performance of the forest department as revealed in chapters 1 and 4. The forest department detected around twenty-five hundred rule violations and convicted about ten thousand persons each year at the peak of the protests by Kumaonis. Even if one takes into account the fourfold population increase that has occurred in Kumaon since the early twentieth century, enforcement by the forest department was far more imperfect than that of the councils. To achieve even its imperfect regulation of environmental practices, the forest department had to increase its size and expenses significantly. The costs of expanding control over thousands of scattered pieces of vegetation were, are, and will be prohibitive for the forest department, especially when villagers contest the expansion of control. The attenuation of forest department ownership has, however, reduced costs. The department's costs were displaced onto the villagers when they began policing themselves.

The second aspect of table 3—the greater the protection effort, the more the rule violations—requires a little explanation but is easy to understand. Violations of existing regimes of allocation also occur in villages where there is no monitoring where councils do not appoint guards, or where guards do not report rule violations. But in these cases council records contain little information about rule violations. Monitoring of ecological practices likely reduces the incidence of violations. But it also brings more violations to light. Thus, it is not surprising to learn that councils that spend large sums on monitoring also document more instances of illegal use. The lack of reported incidents only means that monitoring is lax or nonexistent.

Although forest councils face infractions and rule violations similar to those found by the forest department, they have recourse to a greater variety of monitoring options. The two broad categories into which their options fall are mutual monitoring and third-party monitoring (see figure 7).

Third-party monitoring refers to specialized guard positions that are filled by persons appointed to undertake that task. Under mutual monitoring, there are no specialized guard positions. Within these two

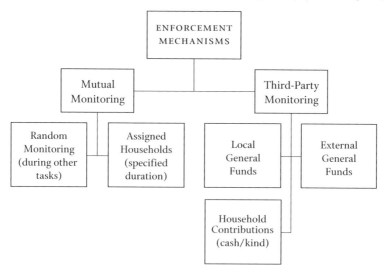

FIGURE 7: Types of monitoring mechanisms employed by Kumaon's forest councils.

broad categories, there can be additional refinements depending on the extent to which villagers are directly involved in monitoring. There is minimal direct involvement under the option in which specialized guards are paid through external funds. Villagers do not have to contribute any funds to support the monitor, and the guard depends for his remuneration on funds from outside the village. The greatest direct involvement of village households occurs when there is no specialized guard position nor specialized monitoring roles. Members of each household monitor all the other households, almost as a by-product of daily life. This form of monitoring is most common in small, close-knit groups.

   The other three forms of monitoring require intermediate levels of villager involvement. As figure 7 shows, mutual monitoring can take two forms. Under the second, households are assigned monitoring responsibilities in rotation. The duration for which they act in the specialized role of a monitor can last anywhere from a day to a month, depending on the size of the village and the extent of its forests. Under third-party monitoring, village residents also have more immediate involvement. When the guard is paid directly by each household through contribu-

TABLE 4.

Distribution of Forest Councils According to Their Forms of Monitoring

Number of Councils	Monitoring Mechanisms				
	Mutual Monitoring		Third-Party Monitoring		
	Random monitoring	Assigned households	Household contributions	Local general funds	External general funds
	2	3	7	18	8

Source: Field research, 1993.

tions in cash or kind, each village household shows a clear commitment to the task of monitoring. The council can pay the guard from funds it has under its control. Typically these funds are raised by the sale of forest products to villagers or from the fines villagers pay when they are caught breaking the rules.

These different forms of monitoring vary in their durability. Thus, the form of monitoring that is the most common and seems to survive best in Kumaon's forest councils is one in which a specialized guard is paid by the council out of its own funds. Different forms of mutual monitoring were prevalent only in a small proportion of the surveyed villages. Table 4 shows the distribution of councils according to the forms of monitoring they have adopted.

The different forms of monitoring are associated with strikingly different beliefs among villagers about forests and the environment (see chapter 6).

One of the most vexing problems in all attempts to supervise guards is the question of how to monitor the monitor.[18] The forest department could not solve it. The Kumaon councils have devised some provisional means to do so, especially in the case of third-party monitoring. Their solution depends on circularity in the monitoring process. Under third-party monitoring, there is a guard who monitors the actions of villagers. The activities of the guard are monitored by the executive committee of the forest council. The executive committee is itself monitored by villagers, thus closing the circle. The relatively small size of the settlements and constant interactions among residents mean that the election of new office holders is as much about finding effective decision makers as it is about making sure that if existing office holders are ineffective they should not be reelected.[19]

Sanctioning

In general, monitoring makes the hidden actions of villagers more visible to power holders. Visibility is the aim of all surveillance. It is also the foundation of subsequent steps to normalize social-ecological behavior. If new regimes of allocation laid down the paths that village residents had to traverse and monitoring made visible deviance and/or conformity to the new paths, forms of sanctioning constituted a new structure of incentives to prompt recalcitrant villagers along desired paths. Forest council officials are a constant presence in the lives of villagers in a manner that the forest department simply cannot match. Their decisions are a reminder of the need to conform to the rules and protect forests. Allocation, monitoring, and sanctioning thus form a connected triad in the institutionalized government of forests.

Monitoring summarizes a number of steps that blend into sanctioning. It involves the patrolling of forests, recording of observed deviance, reporting of deviations to the forest councils, and on occasion direct measures to correct deviance. During their rounds, guards write down the names and actions of anyone they see harvesting products from the forest while prohibitions are in place. They also note evidence of fresh illegal harvesting even if no violator is present. Such instances may involve the presence of an animal in the forest or visual evidence of freshly cut trees, lopped branches, grazed grass, or abandoned cutting tools. When a guard observes an incident of illegal harvesting, he has two options. He can apprehend the culprit, confiscate his or her cutting tools, and report the infraction at the next meeting of the council or, if in doubt, he can simply report the culprit to the council. He maintains a diary in which the names and other particulars of offenders, their specific transgressions, and the details of the event are listed. This information is made available to the council at its next meeting. When appropriate, the guard informs the rule breakers of the date of the next council meeting, at which they must appear to recover their implements and pay their fines.

On being informed, the council decides on the punishment. A variety of mechanisms are available. Offenders may be asked to appear at council meetings to explain their presence and actions in the forest. They can be issued a mild to severe reprimand. They may be asked to render written or public apologies. The council can confiscate cutting implements such as scythes, impound animals, require constructive activities in the forest such as planting or protection, strip offenders of all or some rights, and impose fines. Sometimes these sanctions can have a religious tinge, as when a fine is in the form of an offering required for local deities or gods in the forest. In extreme cases, or when offenders are particularly defiant, the council can exclude them socially, report them to local government officials, invoke the help of higher officials, or seek redress in formal courts.

These sanctions can be seen as elements in an escalating series. For the most part, villagers find the authority of the council a compelling reason to conform. The number of violations at the level of the community may be large, but it is minuscule compared to the level of compliance. When individual village residents violate existing arrangements for using and governing forests, the councils can take increasingly severe steps to ensure that villagers ultimately respect their authority. At each step, the stakes get higher and the number of individuals who need regulation becomes smaller. Most individuals follow the rules. Of the ones whose names are reported by the guard to the council, most alter their actions after being summoned to council meetings to explain why they were in violation of the rules. Reprimands, apologies, confiscation of harvesting equipment, and requirements to undertake constructive activities or provide offerings to local deities reshape the behavior of many others. The imposition of fines is accepted by almost all villagers as being within the jurisdiction of the council, and most pay the relatively small fines within a short period of time. Of those who do not pay the fines initially, many do so after the council threatens them with action for recovery. But a small proportion of villagers is obdurate even after repeated reprimands.[20]

To gain a more textured sense of the dynamics of sanctions, conformity, and noncompliance, consider the case of the Bhagartola forest council. This case also indicates how the burden of sanctions falls unequally on different groups of villagers, even when the sanction is literally in conformance with the idiom of equality.

TABLE 5.

Basic Statistics on the Bhagartola Council Forest

Indicator	Level
Trees per hecatare	1,826.0000
Mean tree DBH[a] (meters)	0.1572
Mean tree height (meters)	6.3000
Total tree biomass (cubic meters per hectare)	205.0000
Number of major tree species	11.0000

Source: Field research, 1993.

[a]DBH is the diameter at basal height.

Bhagartola is located in Almora district in the middle Himalaya, at an altitude of 1,900 meters and is just about 1 kilometer from a paved road. Its forest council was formed in 1937. With seventy households and a forest of 63 hectares, Bhagartola residents have an average of about 1 hectare of forest per household. The village population has not changed much over the past forty years; from 1951 to 1991 it grew from 297 to 328 individuals. The same is true of the goat and cattle population. Between 1961 and 1991, the number of goats was around 150, while the number of cattle fell from 279 to 206.[21] The slow growth of the village population is in part the result of high levels of migration to the plains. Table 5 provides some basic information on the village forest.

The forest is densely vegetated with mixed hardwoods and broad-leaved species such as ainyar (*Andromeda ovalifolia*), kaifal (*Myrica sapida*), rhododendron (*Rhododendron arboreum*), totmila (*Ficus oppositifolia*), and several species of oak (*Quercus sp.*). Villagers depend on the forest and its vegetation for close to 40 percent of their fuelwood and 20 percent of their fodder needs. These products are in short supply, and the Bhagartola forest council has enacted a clear set of rules to limit extraction.

The forest council meets regularly, holding ten meetings a year on the average. In some years, this number rises to fifteen. The minutes of all meetings are recorded and are available to villagers. They are also available to outsiders who can get authorization from revenue or forest department officials. Meetings are devoted to discussions about the state of the forest, the level of rule adherence, specific infractions, and other details connected with the government of the council-managed forest. Two perennial topics of interest are how to raise revenues and enforce rules. The council is always short of funds, and the guard detects villagers breaking rules constantly.

The council has several sources of revenue. The chief ones are payments by villagers for the fodder and fuelwood they harvest, the auction of minor forest products such as fungi and moss, and the sale of pine resin and timber through the Uttar Pradesh Forest Corporation. Revenues from all sources except the last two are readily available to the council to meet its everyday expenses, especially those related to monitoring and protection of the forest. To ensure that rule breaking and illegal harvesting do not reach epidemic proportions, the council summons specific violators to its meetings, strategizes about how to recover fines, and refines the application of rules about extraction of benefits from the forest.

In cases in which individual households are unwilling to listen to the council officials or the guard, the council can seek the services of village-level revenue officials (patwaris) to recover fines. In most instances, the patwari, after some importuning, proves amenable to appeals for help. The successive stages of rule enforcement and the composition of the group of violators at each stage are instructive (see tables 6 and 7).

Tables 6 and 7 contain information for only three years out of a forty-year time span. This renders conclusive statements about trends hazardous. Nonetheless, some important inferences can be made with respect to both tables. In each year, monitoring leads to the detection of a certain number of rule violations. That number seems more or less constant. Assuming a rough proportionality over the years between the number of times villagers break rules and the number of violations the guard detects, it would be fair to suggest that there is a "normal" level of rule violations that cannot be reduced much with existing monitoring technologies and subsistence needs. This level of rule violations arises because there is a mismatch between the significant dependence of villagers on the forest and the desire of the council to prevent forest products from being extracted. But it should be kept in mind, that although villagers break existing rules constantly, the forest is not in a "degraded" condition, as is evident from personal observations and in the measurements of vegetation reported in table 5.

Second, the social identity of the offenders is not in proportion to

TABLE 6.

Gender Composition of the Village Population and Rule Violations

	I Number of Adult Individuals in the Village	II Offences Reported to Council	III Cases in which Fines Not Paid Initially	IV Cases in which Fines Not Paid for at Least a Year
1951				
Men	33 (49)	38 (19)	20 (36)	8 (58)
Women	35 (51)	163 (81)	38 (64)	6 (42)
Total	68	201	58	14
1971				
Men	30 (48)	37 (17)	23 (47)	6 (67)
Women	33 (52)	181 (83)	25 (53)	3 (33)
Total	63	218	48	9
1991				
Men	38 (47)	50 (27)	39 (65)	18 (78)
Women	41 (53)	135 (73)	21 (35)	5 (22)
Total	79	185	60	23

Source: Bhagartola forest council, records of meetings in 1951, 1971, and 1991.

Note: Figures in parentheses are proportions of the total (population/rule violations).

their population in the village. Consider column II in table 6, for instance. The figures in parentheses represent the proportions of total rule violations committed by men and women. The proportion of detected violations by women is higher than their proportion in the village population. This might in part reflect the fact that it is women who are primarily responsible for collecting fodder and fuelwood for the household.

We come now to table 7. Brahmans enjoy a higher ritual status than thakurs. The thakurs, or the rajputs, are possibly stronger politically than the other two castes, but the harijans are socially inferior to both brahmans and thakurs. In the absence of information on landholding and wealth, caste can be taken as a proxy for social inferiority and lack of power. In table 7, the disproportionality for caste is similar to that found for gender in table 6. The number and proportion of detected offenses by harijans are far higher than those for brahmans and thakurs. The higher proportion can be explained by the common observa-

TABLE 7.

Caste Composition of the Village Population and Rule Violations

	I Number of Adult Individuals in the Village	II Offences Reported to Council	III Cases in which Fines Not Paid Initially	IV Cases in which Fines Not Paid for at Least a Year
1951				
Brahmans	45 (66)	71 (35)	28 (49)	8 (57)
Thakurs	5 (7)	10 (5)	11 (19)	4 (28)
Harijans	18 (27)	120 (60)	19 (32)	2 (15)
Total	68	201	58	14
1971				
Brahmans	43 (68)	88 (40)	25 (52)	5 (55)
Thakurs	5 (8)	16 (7)	14 (29)	3 (33)
Harijans	15 (24)	116 (53)	9 (19)	1 (11)
Total	63	220	48	9
1991				
Brahmans	49 (60)	55 (30)	42 (71)	17 (73)
Thakurs	6 (8)	16 (9)	10 (16)	5 (22)
Harijans	25 (32)	114 (62)	8 (13)	1 (5)
Total	80	185	60	23

Source: Bhagartola forest council records of meetings in 1951, 1971, and 1991.

Note: Figures in parentheses are proportions of the total (population/rule violations).

tion that poor villagers are often more dependent on common property resources than are those with more private assets (Jodha 1990), even in the Indian middle hills, where social stratification is less striking than in the plains.

Consider columns III and IV in the two tables. The proportion of those who pay their fines is far higher for women and harijans than for men and upper-caste individuals. It seems that the council is less able to enforce its writ among upper-caste individuals and men than it is among women and the harijans. Far more of the women pay the fines imposed on them than do men. And the proportion of brahmans who conform to the sanctions that are imposed on them is far smaller than that of the harijans. The rules of the forest council apply to all village residents equally. But not all villagers follow them equally, nor can the council enforce them equally. Women depend more on the forest because of social norms that make them responsible for household chores. Women also feel compelled to a greater degree to pay their fines. The point here is not just that guards are stricter in enforcing rules against women, although that may also be true.[22] It is that those who have an inferior status in the village and are not as strong politically or socially, including the harijans, feel a greater imperative to follow rules.

In reporting information on compliance and defiance by gender and caste, I do not mean to subscribe to any static or stable notion of these terms. Nor do I mean to read local politics off these categories directly. And, finally, I do not mean to suggest that persons who defy existing patterns of expectations and compliance will continue to do so. Indeed, the next chapter examines some of the conditions under which individuals and households come to accept and defend new forms of behavior and institutions. It also examines how a politics of subjecthood unfolds rather than using membership in a social category to define the nature of the subject. I recognize that there are many local variations within the categories brahman, thakur, and harijan or men and women. I also recognize that the political status of those belonging to particular categories is not necessarily determined by such belonging. It is crisscrossed by marriage and kinship relations, occupational status, relations to the state, and, most importantly, involvement in different social practices. My use of these categories here serves a limited purpose, different from the larger questions of how to understand social categories such as caste and gender or conceptual categories such as identity and subjecthood. The point is simply that regulatory rules, even when they are seemingly equitable, can produce outcomes that are systematically biased against those who are marginal and less powerful. To understand the effects of regulation, therefore, it is important to focus not just on their formal-literal meaning but also on the sociocultural and the political-economic context in which they are (selectively) enforced. Given the intrahousehold division of labor, women are more dependent on forests than men are. They feel more constrained to follow the arrangements that the council has created. Harijans are less powerful than brahmans and thakurs. They also feel they must conform to council rules to ensure continued access to forest products.

If institutions are seen as rules intended to generate particular patterns of behavior, then it is critical to pay attention to those unwritten norms that influence behavior implicitly, and perhaps just as systematically, as the rules that are written and explicit. The character of sanctions in Bhagartola is determined by prevailing social norms. These norms are also an index to power. As the unwritten rules of the game, they permit upper-caste villagers to get away with far higher levels of nonconformity than the lower-caste households can. These unwritten rules are neither explicitly negotiated nor equitable. They are, rather, the reflection of the structured exclusion from power to which some of the villagers are subject. In the Bhagartola example, lapses in rule enforcement have a particular bias against women and lower-caste members. There is little reason to believe that Bhagartola is an exceptional case in Kumaon. The community treats some of its members more equally than others, even as the contractual underpinnings of the regulatory community claim to cast them as formal equals. This is not to suggest, of course, that only recent, state-facilitated communities create unequal rules or enforcement patterns. The lattha panchayats that predated the forest councils of today depended on more locally defined sources of power for their existence, but their power was likely also exercised unequally in allocating and regulating forests.[23]

Adjudication

If a local resident does not conform to council regulations, such challenges to the council's authority can be addressed in two ways. Councils can appeal to the local revenue official (the patwari) or higher level revenue department authorities. In extreme cases, a council can take a recalcitrant villager to court. Neither option is very satisfactory. Revenue officials are burdened with tasks in their own departments, which makes their responses slow. Trying to use the overburdened and cumbersome formal judicial system to recover the small sums that an offender typically owes the council is tantamount to spending a fortune to get a nickel back. Typically councils resort to the court system only when a villager has encroached on the forest land or illegally felled a large number of trees.

The Bhagartola forest council has never had to appeal to the courts. In fact, few councils ever do. Their authority includes some powers of adjudication. They decide on the guilt of villagers the guard detects, impose sanctions, and act as arbiters when there are disagreements over sanctions and interpretations of allocation and monitoring rules. They can reduce or lift sanctions imposed earlier, resolve disputes between villagers and guards, and change the dates and duration for which forest compartments are made available to villagers.

Villagers are willing to accept these judgments because of the structural asymmetry between their power and that of the council. If the council does choose to take a villager to court, its word is likely to carry greater weight. The information about repeated violations that would be available in the council's notes of its meetings, the testimony of a number of council officials against the recalcitrant villager, and the word of the guard are all likely to persuade the court to decide in favor of the council's position. Further, the council will likely have more funds with which to contest a case in court. The weight of these facts suggests that almost all users are unwilling to test the resolve of a council to the point of going to court. It is in the villagers' interests not to have to go to court. It is this ultimate loading of the dice in favor of the council that drives intermediate outcomes in its favor. Therefore, when villagers are apprehended in the forest by the guard, they are willing to appear before the council, render apologies, promise not to break rules again, and pay their fines.

The Characteristics of Regulatory Strategies: Community versus the Forest Department

The chief objective of regulation in Kumaon, as shaped by the means of community, is adjustment and guidance of villager actions in terms of quantity, force, space, time, and predictability. No longer can villagers extract from their forests the products they need, when they want, and in the manner they consider appropriate from the locations that prove convenient or in the quantities necessary for the household. Kumaon's councils create rules to cover all of these contingencies and others. They enforce their regulations with all the resources they can muster. The new instrument of social-ecological engineering that the councils have become is more intimate, minute, daring, complete, and comprehensive than the forest department ever was. The department tried to eliminate villager presence and actions in the forest altogether. Although villagers could not be completely prevented from entering and using the forest, it was at least necessary to change the nature and character of their actions.

Decision makers in the community use detailed regulations as the means to create a new economy of villagers' use of forests, one that is dependent for its effectiveness as much on the sociality that exists in the village as on the sheer calculus of gains and losses. The attempt to adjust and guide, one might say, is no more than the recognition of a necessity. In trying to eliminate human interventions completely, the forest department aimed at too broad a target. Its severe actions missed their mark as often as they found it. By drawing back from such an encompassing goal and making communities the bearers of regulatory strategies, the state simultaneously redefined the objective of regulation: from exclusion and prohibition to normalization and adjustment.

The department itself was ill-suited for the achievement of such an objective. Its staff was too small, the area under its control was too large, the organization of its activities was too hierarchical, and the sources of resistance to its strategies were too prolific. Attempts to normalize require a conception of the desirable, an intimate knowledge of the object of regulation, the ability to calculate the effects of policing strategies, the capacity to distinguish between human and other causes, and the flexibility to modulate the nature and force of interventions. The use of statistics and numbers permitted the department to subject vegetation to new regulatory strategies and learn the effects of its strategies over time, although even here its knowledge was no better than imperfect. But when it came to human actions and their modulation the forest department discovered that it had much less room to maneuver.

In the realm of allocation of products, its hands were tied because of the overriding compulsion to make profits. Timber was the most valuable forest product commercially, and it was at maximizing the production of salable timber that the department aimed. It was too costly to devise and enforce allocation schemes for other products. These relationships were visible in the annual reports the department published. Revenues from timber far exceeded those from all other products put together. No wonder the department preferred to exclude rather than regulate, prohibit rather than modulate. The pursuit of higher revenues and profits led the department to proscribe villagers from using forests classified as reserves and prohibiting actions such as lopping, firing, and grazing.

A similar grossness characterized the department's ability to gather information. Its knowledge of what villagers did in the forest was fated to be filtered through knowledge categories related to the production of timber. Did fire, grazing, firewood collection, and fodder harvests affect the production of wood adversely? Could villagers gather nontimber forest products without harming regeneration? Could the legal actions of villagers be separated from their illegal actions? Depending on the answers to these questions, the department would formulate its exclusionist strategies. It was easier to enact new prohibitions than to seek accurate answers to these questions given the spatial patchiness and heterogeneity of the answers. The department's knowledge of what the villagers actually did, its ability to monitor and gather information on the effects of its actions, and its capacity to modulate its strategies were limited at best. An enormous amount of illicit behavior went undetected; indeed, it was undetectable. Questions about the effects of fire and lopping could not be settled in any satisfactory manner. They continue, even today, to be subjects of debate among silviculturists and ecologists.

Even when the department came to know what the effects of its proscriptions and interventions were, the range of instruments it could deploy as sanctions was minimal. To improve the effectiveness of monitoring, the only practical alternative for the department was to hire more guards. But guards were prone to intimidating poor villagers and were vulnerable to bribes from timber merchants and contractors. The presence of more guards created the dilemma of how to monitor the monitor. And when it came to sanctions the department's favored remedies were court cases, fines, and prison sentences. The conviction rate for cases that the department took to court was typically in excess of 90 percent. The department preferred prison sentences for those it apprehended in the forest. Villagers were usually too poor to pay fines that would offset the costs of monitoring and legal fees incurred in court. Imprisonment at least removed the risk that the person would break the rules again.

These remedies were too sweeping and alienating, only reinforcing the cycle of violations, sanctions, resentment, and further violations. Where villagers and the department relied on the same resource—one for their livelihoods and the other for revenue—the ensuing conflicts were too widespread to be contained under the existing strategies.

The strategies adopted by decision makers in communities are different, most obviously in the range they span. The forest councils are concerned not just with the extraction of timber but also with the harvesting of green manure, fodder, leaves, firewood, nontimber forest products, and construction materials. For each of these products, they craft different regimes of harvesting and use. Their interest in shaping and regulating how villagers extract these goods is backed by the use of different forms of monitoring. Their monitoring mechanisms require varying levels of participation from the villagers and have different levels of associated costs.

Often the council combines various monitoring mechanisms. Through assiduous and consistent monitoring, councils are aware of a significant proportion of villagers' actions in the forest. If the forest department detected less than even 1 percent of what happened in the forests in violation of the use and management regimes it had created, the councils detect closer to 20 percent. Their records and documentation make villagers' actions known to a larger audience, including state officials and other villagers. In bringing visibility to actions hidden in the forest, the councils have enlarged the sphere of what regulatory strategies can influence and adjust. One of the most important effects of the different monitoring strategies is changes in villagers' understandings of the environment and their relationship to environmental protection—a theme that the next chapter develops further.

The significant range of monitoring instruments is matched by the ability of the councils to choose different forms of punishment to suit the illicit actions they detect. Unlike the forest department, whose ability to deter unwanted actions in the forest rested on the efficacy of the twin instruments of fines and imprisonment, the council has access to a wider and finer range of sanctioning mechanisms. What Ostrom (1990: 94–100) calls "graduated sanctions" differ not only in their levels and strengths but also in their types. Councils use sanctions that can affect their targets socially, economically, and politically or in some combination of these dimensions. The choice of a particular penalty depends on any number of factors, among them the identity of the offender; the nature, seriousness, context, and frequency of the infraction; the history of interactions between the council and the offender; and the strength of the council. For the forest department, it is well nigh impossible either to collect the information necessary to deploy these different types of sanctions at appropriate levels or to decide which sanction will have the greatest effect. The ability of the councils to detect a larger proportion of infractions and apply penalties with finer discernment serves as a strong support for the arrangements they create to regulate the use of forests.

The consistent choice of forms of allocation that suit villagers' needs, mechanisms of monitoring that balance costs and participation, and types of sanctions that match the gravity of the villagers' actions grants the councils' regulatory actions far greater acceptability than the forest department could gain. The councils' voluminous records on incomes and expenses, minutes of meetings, visits of forest council inspectors, means employed to protect forests and monitor use, rule infractions and the identity of those who break rules, sanctions imposed on villagers, and variations in these over time are powerful instruments that make local practices visible. Once brought into existence, these means of uncovering and exposing actions to the eye, even by themselves, discourage forest users from harvesting fodder, firewood, or timber in violation of prescribed levels.

It is only fair to point out that councils are not always an effective instrument of regulation. In the realm of allocation, their actions may be far from successful, especially when the community is extremely stratified, regeneration in the forest is meager, and the dependence of villagers on the forest is limited. Their efforts to monitor effectively are hindered in the absence of sufficient funds with which to hire monitors and pay their salaries. Their ability to sanction those who do not obey rules depends significantly, even if only partially, on the extent to which government officials in the revenue and forest departments are willing to lend support.

Thus, to highlight the differences in the abilities of the forest department and the councils is not to suggest that the councils have displaced the forest department. Rather, it is to document the emergence and delineation of a new government of the environment. It is to show how the councils can undertake certain regulatory tasks more precisely, directly, economically, and continuously. It is to highlight how the councils and state officials can work in tandem, the capacities of each being complementary. Deficiencies in certain arenas of allocation and enforcement led the forest department to delegate some of its tasks to the community. The forest councils sharpened and targeted the processes

of regulation that had overwhelmed the department. Their empowerment permitted official power to penetrate locations that it had earlier
been unable to colonize.

The 3,000 square kilometers of forests that Kumaon's villagers have
come to control are the basis of their livelihoods. For many villagers,
survival would be a more difficult proposition in their absence. In the
absence of massive protests at the beginning of the century, these forests would likely have been incorporated and consolidated into the
centralized system of allocation, monitoring, and sanctions that the
forest department represents. The new governing agency for this territory is a hybrid born of members of the state and the community. It
comprises the forest councils and their leadership together with state
officials such as forest council officers and inspectors, forest guards,
patwaris, and divisional forest officers. This new body of allocation and
regulation can be seen as a mediating layer between the marginal villagers, who interact with state officials relatively infrequently, and
other, more powerful officials who visit villagers rarely, if at all.

The efficacy of the new regulatory regime in Kumaon thus depends
on a marriage between the diversity of the forest councils and the
delegated power of state officials. The councils' monitoring and sanctioning abilities and the power of state officials combine to make the
new regulatory regime in Kumaon a powerful instrument for reconfiguring resource use and users' views. The foundation of this regime is
the recognition of its members that they cannot by themselves accomplish the project of government. Its comprehensive reach stems from
yoking together two social actors that had seen their interests as divergent but have now come to accept their mutuality and interdepen-
dence. The legitimacy of the regime derives not from collective visions
of dazzling projects of sustainable development and large profits but
from the tempting promise that if villagers restrain their current consumption levels their needs will be met indefinitely.

Regulatory Rule in Communities

At stake in the discussion of regulatory rule in Kumaon is the birth of a
new form of government based on strategies that redefine the interests
of multiple actors, join them along new axes of common purpose, and

enforce the protection of environment more economically in the process. Regulatory rule by communities is different from more familiar models of crime and sanction, infraction and deterrence, and violation and punishment. For example, in exploring and explicating transformations in the nature and forms of penality, Foucault invokes two metaphors: the punitive city and the more familiar mechanism of Bentham's panopticon. The punitive city for Foucault is a possibility that ultimately cannot be realized. He describes it as follows.

> This, then, is how one must imagine the punitive city. At the crossroads, in the gardens, at the side of roads being repaired or bridges built, in workshops open to all, in the depths of mines that may be visited, will be hundreds of tiny theatres of punishment. Each crime will have its law; each criminal his punishment. It will be a visible punishment, a punishment that tells all, that explains, justifies itself, convicts: placards, different coloured caps bearing inscriptions, posters, symbols, texts read or printed, tirelessly repeat the code. . . . The essential point, in all these real or imagined severities, is that they should all, according to a strict economy, teach a lesson: that each punishment should be a fable . . . in counterpoint with all the direct examples of virtue, one may at each moment encounter, as a living spectacle, the misfortunes of vice . . . and popular memory will reproduce in rumour the austere discourse of law. ([1975] 1979: 113)

The lessons of the punitive city operate in a quite specific fashion. They make the concrete relationship between a crime and its punishment obvious through public examples. They produce in their beholders the awareness of what will happen *before* a crime is committed. They create and activate a mental relationship between criminality and its just desserts so that the very thought of infringing rules may be vanquished. The discipline of the punitive city does not depend on establishing relations of visibility between the criminal and the supervisor. Government does not require that sanctions be implemented consequent to the committing of a crime. Rather, those who would engage in extralegal acts are given the knowledge of the effects of such acts well in advance of the acts themselves. Through aversion to consequences does the punitive city function.

In contrast, the panopticon reverses the temporal relationship between criminality and its outcomes. The individual who would break the law must reckon with the consequences of his or her actions after committing the crime. Consider how Foucault describes this coercive institution, with Bentham as his source.

> At the periphery an annular building; at the centre, a tower with wide windows that open onto the inner side of the ring; the peripheric building is divided into cells, each of which extends the whole width of the building; they have two windows, one on the inside, corresponding to the windows of the tower; the other on the outside, allows the light to cross the cell. . . . All that is needed, then, is to place a supervisor in a central tower and to shut up in each cell a madman, a patient, a condemned man, a worker or a schoolboy. . . . The panoptic mechanism arranges spatial unities that make it possible to see constantly and recognize immediately . . . [that] visibility is a trap. . . . Hence the major effect of the Panopticon: to induce in the inmate a state of conscious and permanent visibility that assures the automatic functioning of power. . . . It automatizes and disindividualizes power . . . [it] produces homogeneous effects of power. . . . He who is subjected to a field of visibility, and who knows it . . . becomes the principle of his own subjection. ([1975] 1979: 200–203)[24]

Whatever one might think of the automatic relation between visibility and normalization that Foucault asserts to be active in the panoptic mechanism, it is obvious that regulatory rule is different from the punitive city and the coercive institution in its operations as well as its effects. The punitive city and the coercive institution both position the rule violator on the outside. Regulatory rule in Kumaon, at least in the first instance, appeals to a sense of membership within the community where the rule violators reside. The temporal relationship between committing a rule infraction and suffering its consequences is also different under regulatory rule. Rather than producing discipline by invoking an a priori relationship between crime and sanctions, or by establishing an a posteriori connection between the gaze of power and the normalization of the individual, regulation works both before and after infractions. Members of a community are aware of the nature of sanctions and the quality of monitoring because they themselves participate in creating and enforcing them.

Further, the forms of penality that Foucault discusses depend on direct relations of visibility: the loud and spectacular displays of retribution under sovereignty, the observable relationship between crime and its return effects in the punitive city, and the relentless gaze of power associated with the panopticon. Power works through and along the lines of observation that are made possible by sight. It is absolute in its effects. Those who are exposed to the lessons of crime in theaters of punishment learn the effects of deviance and are scared away from it. Those who are subjected to the gaze of power undergo changes in their subject positions. Those who are not a part of the audience in the theater or who are not subjected to the gaze continue to act in ways that power would sanction.

In comparison, regulatory rule creates awareness and knowledge through direct participation in the various elements and stages of regulation. Those who take part in allocating resources, monitoring actions in forests, and implementing sanctions are more likely to come to appreciate the fragility of the environmental resources they are trying to conserve. Those who see the environment as requiring protection are more likely to put greater effort in their protectionist practices. Involvement in the practices of rule making and the conduct of rule enforcement becomes effective not just because of the presence of mechanisms of visibility and lines of observation. Involvement generates awareness and knowledge by confronting subjects with the effects of their actions as they undertake them. The effects of this form of regulation are cumulative over time rather than absolute in the first instance. Discovery of deviance is uncertain at any given moment. But over time the probability of being apprehended increases if those subject to governmental strategies of regulatory communities continue to violate prescribed norms. Transformations in subject positions, as the next chapter will describe, are similarly probabilistic. Involvement in regulatory practices and awareness of collective decisions contribute to shifts in environmental practices as well as beliefs. The regulatory community relies for its effectiveness on the rules of chance rather than the absolutism of vision. The shift from the dynamic of coercion and resistance toward one of involvement in regulatory practices and transformations in environmental subjectivities may be an uncertain process, but it is the goal toward which regulatory communities strive.

6

Making Environmental Subjects:

Intimate Government

Honest, sober, frugal, patient under fatigue and privations, hospitable, good humored, open, and usually sincere in their address, [Kumaonis] are at the same time extremely indolent, fickle, easily led away by the counsel of others, hasty in pursuing the dictates of passion, envious of each other, jealous of strangers, capable of equivocation and petty cunning, and lastly, grossly superstitious.—Raper, quoted in Atkinson, *The Himalayan Gazetteer*

Down the street an ambulance has come to rescue an old man who is slowly losing his life. Not many can see that he is already becoming the backyard tree he has tended for years.—Joy Harjo, *Transformations* (*How We Become Humans*)

Recall Hukam Singh from chapter 1. He started out an environmental skeptic. But over the period I knew him his concern for the environment grew to a point where he came actively to defend the need for environmental protection and regulation. He found efforts by the forest department wanting and saw villagers as the real basis for appropriate forms of regulation. The shifts in Hukam's beliefs hint at what is perhaps the most important and underexplored question in relation to environmental regulation: when and for what reason do socially situated actors come to care for, act, and think of their actions in relation to something they define as the environment? Hukam Singh did not care much about the village forest in 1985, but he was defending the need for its regulation by 1993. Similarly, concern for the environment has not always existed in Kumaon. It has emerged and grown over time. Widespread involvement in specific regulatory practices is tightly linked to the emergence of a greater concern for the environment and the creation of environmental subjects. I use the term *environmental subjects* to nominate those who thus care about the environment. More precisely, the environment constitutes for them a conceptual category that organizes some of their thinking; it is also a domain in conscious relation to which they perform some of their actions. The practices and thoughts of environmental subjects, as I define the term, may not always lead to environmental conservation. But they are often undertaken in relation to the environment. I draw on evidence related to forests as an example of an environmental resource. Further, in considering an actor as an environmental subject I do not demand a purist's version of the environment: something necessarily separate from and independent of concerns about material interests and everyday practices. A desire to protect commonly owned or managed trees and forests, even with the recognition that such protection could enhance one's material self-interest, subscribes to environmental subjectivities. In such situations, self-interest comes to be cognized and realized in terms of the environment.

If the environmental aspect of *environmental subjects* requires some boundary work, so does the second part of the phrase.[1] It should be evident that I do not use *subjects* in opposition to either *citizens* or *objects*. One commonsense meaning of *subjects* would be to see them as actors or agents. But subjected people are also subordinated. The third obvious referent of the term is the notion of a theme, or a domain, as in the environment being the subject of my research (see chapter 7 for a discussion of Foucauldian meanings of subjection). I use subjects to think about Kumaon's residents and changes in their ways of looking at, thinking about, and acting in forested environments in part because of the productive ambiguities associated with the term *subject*. Each of its referents is an important part of the explorations in this book.

Given the existence of environmental subjects in Kumaon, a second question can be asked: what distinguishes them from those who still do not care or act in relation to the environment? Of the various residents of Kotuli, only some have changed their beliefs about the need for forest protection. Others neither take an interest nor work in relation to protectionist regulations. They harvest forest products without attending to locally formulated enforcement. The story is more widely applicable in Kumaon. Thus, to say that Kumaonis have come to care about their forests and the environment is only to suggest that some of them, in increasing numbers over the past few decades perhaps, have come so to care.

Answers to questions about who acts and thinks about the environment as a relevant referential category and when, how, and why one so acts are the quarry that this chapter seeks to hunt. These answers are important for both practical and theoretical reasons. Depending on the degree to which individuals care about the environment, the ease with which they will agree to contribute to environmental protection may be greater and the costs of enforcing new environmental regulations may be lower. But equally important, and perhaps even more so, is the theoretical puzzle. What makes certain kinds of subjects, and what is the best way to understand the relationship between actions and subjectivities? Against the common presumption that actions follow beliefs, I suggest, using evidence from Kumaon, that people often first come to act in response to what they see as compulsion or in their short-term interest and only later develop beliefs that defend short-term actions on other grounds as well. Against the general tendency to explain environmental outcomes in terms of static social categories such as caste, gender, and wealth, I highlight the role of practice and regulation. I will show how residents of Kumaon differ in their beliefs about forest protection, and how these differences are related to their involvement in regulatory practices.

It can be argued that beliefs and thoughts are formulated in response to experiences and outcomes over many of which consciously formulated strategies by a single agent have little control. Individuals strategize about the future and attempt to shape the contexts of their interactions. There can also be little doubt that one can change some aspects of the world with which one interacts. But equally certainly, the number and types of forces that affect even one's daily experiences transcend one's own will and design. Much of what one encounters in the world is only partly a result of planned strategies undertaken in light of one's existing beliefs and preferences.

The underlying argument is straightforward. At any given moment, people may plan to act in accordance with their existing preferences. But all plans are incomplete and imperfect, and none incorporates the entire contextual structure in which actions lead to consequences. For these and other reasons, actions lead to unanticipated outcomes. The experience of these unanticipated outcomes does not always harden actors in their existing understandings. Some outcomes may show how earlier subject positions were mistaken. In these situations, actors have an incentive to reconsider existing preferences and subjectivities, incorporating into their mentalities new propensities to act and think about the world. Even if only a very small proportion of one's daily experiences undermine existing understandings, over a relatively short period there may be ample opportunities to arrive at subject positions quite different from those held earlier.

In any event, persuasive answers about variations between subject positions and the making of subjects are likely to hinge on explanations that systematically connect policy to perceptions, government to subjectivity, and institutions to identities. Environmental practice, this chapter suggests, is the key link between the regulatory rule that government is all about and imaginations that characterize particular subjects. In contrast, social identities such as gender and caste play only a small role in shaping beliefs about what one considers to be appropriate environmental actions. This is not surprising. After all, the politics of identity considers significant the external signs of belonging rather than the tissue of contingent practices that may cross categorical affiliations. In the subsequent discussion, I hope to sketch the directions in which a satisfactory explanation needs to move.

Producing Subjects

The description of my meetings and conversations with Hukam Singh, although it seems to be located quite firmly in an overall argument about the emergence of new subjectivities, albeit in relation to the environment, resembles Geertz's idea of "a note in a bottle." It comes from "somewhere else." It is empirical rather than a philosopher's "artificial" story, and yet it has only a passing relationship to representativeness (Greenblatt 1999: 14–16). Connecting it to a social ground and other roughly similar stories requires the development of some crucial terms and the presentation of additional evidence. Two such terms are *imagination* and *resistance*.

In his seminal account of the origins of nationalism, Anderson famously suggests that the nation is an imagined community ([1983] 1991). In a virtuoso performance, he strings together historical vignettes about the development of nationalism in Russia, England, and Japan in the nineteenth century (83–111) to show how these cases offered models that could successfully be pirated by other states, where "the ruling classes or leading elements in them felt threatened by the world-wide spread of the nationally-imagined community" (99). The model that, according to Anderson, comes to triumph is that of "official nationalism."[2] He suggests that official nationalisms were:

> *responses* by power groups . . . [which were] threatened with exclusion from, or marginalization in, popularly imagined communities. . . . Such official nationalisms were conservative, not to say reactionary, *policies* . . . Very similar policies were pursued by the same sorts of groups in the vast Asian and African territories subjected in the course of the nineteenth century . . . [and] they were [also] picked up and imitated by indigenous ruling groups in those few zones (among them Japan and Siam) which escaped direct subjection. (110)

It is an interesting and disturbing fact that for Anderson the successful adoption, superimposition, and spread of official nationalism as a substitute for popular nationalism lay well within the capacity of ruling groups to accomplish, despite the imagined nature of nationalism. A number of scholars have elaborated on the term *imagination*, imaginatively, in writing about the nation (e.g., Appadurai 1996: 114–15; and Chakrabarti 2000a: chapter 6). But in *Imagined Communities* itself, the subsequent analysis gives relatively short shrift to several aspects of imagination. The successful imposition of an official version of nationalism around the globe, coupled with the imagined quality of national emergence that is the core of Anderson's intervention, implies that power groups were able to colonize the very imaginations of the people over whom they sought to continue their rule. They did so in the face of existing popular nationalisms. How they overcame, even for a few decades and certainly only patchily, the resistance that existing senses of "imagined belonging" posed to their efforts requires further elaboration than Anderson provides. The politics at the level of the subject that is likely involved in the struggle between official and popular nationalisms needs additional work if it is to be compellingly articulated.[3] National subjects, to use a shorthand term to refer to the colonization of political imagination by official nationalizing policies, emerged in history. A history of nationalism therefore needs a politics of the subject.[4]

The question about when, why, and how some subjects rather than others come to have environmental consciousness is precisely similar to what Anderson leaves out when he considers the nation. Similar judgments about the transformation of the consciousness of those who are less powerful can be found in the work of other scholars. Consider Barrrington Moore, who writes: "People are evidently inclined to grant legitimacy to anything that is or seems inevitable no matter how painful it may be. Otherwise the pain might be intolerable" (1978: 459). One might ask Moore: "All people?" If not all, then surely we are forced to consider the questions of which ones, when, why, and how? The same motivation to account for social and political acquiescence impels Gaventa's brilliant study of power and quiescence in Appalachia (1982). But his analysis of the third face of power can be supplemented by the examination of mechanisms that would explain when and how it is that some people come to accept the interests of dominant classes as their own *and others do not.*

In contrast to Anderson, for whom the imagination of the less powerful subject is unproblematically appropriable by official policies, scholars of resistance have often assumed the opposite. For them, the resisting subject is able to protect his or her consciousness from the colonizing effects of elite policies, dominant cultures, and hegemonic ideologies. This ground truth forms both their starting assumption and their object of demonstration. Scott's path-breaking study of peasant resistance (1985), his more general reflections on the relationship between domination and resistance (1990), and the work on resistance that emerged as a cross-disciplinary subfield in the wake of his interventions have helped make familiar the idea that people can resist state policies, elite power, and dominant ideologies. Scott assertively advances the thesis that the weak withstand the powerful, perhaps always, at least in the realm of ideas and beliefs. He also suggests that when signs of their autonomous views about the prevailing social order are invisible it is because of material constraints and their fear of reprisal upon discovery, not because they have come wholeheartedly to acquiesce in their own domination, let alone because their consciousness has been incorporated into a hegemonic ideology.

Scott articulates this position most fully, perhaps. But a similar understanding of peasants and their interests was also complicit in the early efforts of subaltern studies scholars to identify an autonomous consciousness for the excluded agents of history.[5] Consider, for example, Ranajit Guha's (1982a) seminal statement on the historiography of colonial India. This powerful manifesto against elitist history writing, in calling for a more serious consideration of the *"politics of the people,"*

portrays the subaltern as "*autonomous*" and subaltern politics as structurally and qualitatively different from elite politics in the sense that
"vast areas in the life and consciousness of the people were never integrated into [bourgeois] hegemony."[6] Even those who note that the opposition between domination and resistance is too mechanical to cap-
ture how the consciousness of those subjected to power changes in
their experience of power go on simply to note that the process is
"murky" (Comaroff and Comaroff 1989: 269, 290). But for scholars of
resistance and subalternity, the autonomous consciousness of peasants,
subalterns, and other marginalized groups endures in the face of dominant elite pressures operating in a spectrum of domains, not just in the
domain of policy. It is reasonable to infer that if for these scholars the
weak and the powerless can resist the panoply of instruments available
to dominant groups policy by itself is even less likely to affect the
consciousness of the subject.[7]

It is clear that the works discussed above present two aspects of the
puzzle about the relationship between government and subjectivity.
Each of these aspects comprises strong arguments in favor of a particular tendency: in one case, the tendency toward the colonization of the
imagination by powerful political beliefs; and in the other the tendency
toward durability of a sovereign consciousness founded on the bedrock
of individual or class interest. Within themselves, these arguments are
at least consistent. Considered jointly as a potential guide to the relationship between the subject and the social they lead to conflicting
conclusions. It is crucial not to account just for the persistence of a
certain conception of interests among a group of people nor just to
assume the straightforward transformation of one conception of interests into another but to explore more fully the mechanisms that can
account for various possible effects on people's conceptions of their
interests (Wildavsky 1987, 1994).

I weave a path through the opposed conclusions of two different
streams of scholarship by suggesting that technologies of government
produce their effects by generating a politics of the subject that can be
understood and analyzed better by considering both practice and imagination as critical. The reliance on imagination by some scholars (Appadurai 1996; Chakrabarty 2000a) in thinking about the emergence of
different kinds of subjects is a step in the right direction. But closer
attention to social practices can lead to a species of theorizing that

would be more tightly connected to the social ground where imagination is always born and, reciprocally, which imagination always influences. A more direct examination of the heterogeneous practices that policy produces, and their relationship with varying social locations, has the potential to lead analysis toward the mechanisms involved in producing differences in the way subjects imagine themselves. My interest is to highlight how it might be possible and why it is necessary to politicize both community and imagination in the search for a better way to think about environmental politics.

Foucault's insights on the "subject" form a crucial point of reference but also a point of departure in considering the political that is silenced in Anderson's vision of the imagined community. In *Discipline and Punish*, Foucault elaborates a particular model of subject making in which "bodies, surfaces, lights" are so arranged as to facilitate the application of power in the form of a gaze. "He who is subjected to a field of visibility, and who knows it," he writes, "assumes responsibility for the constraints of power; he makes them play spontaneously upon himself; he inscribes in himself the power relation in which he simultaneously plays both roles; he becomes the principle of his own subjection" ([1975] 1979: 202–3). The panopticon, standing for such arrangements of bodies, surfaces, and lights, does not just create lines of visibility; it can also be used as a machine "to train or correct individuals" and a laboratory "to perfect the exercise of power" (205–6). Here, then, is a mechanism—the gaze—that acts as a sorting device. Those subject to the gaze become subject to power, examples of the effects it can produce. Those who escape the gaze also escape the effects of power.

Although this example introduces political practice into the process by which subjects make themselves, it will obviously not do. By itself, the model needs more work for any number of reasons, among them its absence, even in total institutions, and the unfeasibility of applying its principles outside them.[8] Nor is it the case that visibility in asymmetric political relationships necessarily produces subjects who make themselves in ways desired by the gaze of power. And, although Foucault does not elaborate on the specific mechanisms that are implicated in the making of subjects (Butler 1997: 2), he recognizes the indeterminacy inherent in the making of subjects when he suggests that mechanisms of repression can produce both subjugation and resistance (1978: 115). It is this recognition of contingency that introduces the register of the political in the creation of the subject. It is also precisely what Appadurai has in mind when he suggests that colonial technologies left an indelible mark on the Indian political consciousness but that there is no easy generalization about how and the extent to which they "made inroads into the practical consciousness of colonial subjects in India." Among the dimensions he mentions as important are gender, distance from the colonial gaze, involvement with various policies, and distance from the bureaucratic apparatus (1996: 134).

These factors are important, of course. But it is necessary to make a distinction between the politics generated by involvement in different kinds of practices and the politics that depends on stable interests presumed to flow from belonging to particular identity categories. Much analysis of social phenomena takes interests as naturally related to particular social groupings: ethnic formations, gendered divisions, class-based stratification, caste-categories, and so forth. Imputing interests in this fashion to members of a particular group is common to streams of scholarship that are often seen as belonging to opposed camps (Bates 1981; Ferguson [1990] 1994). But doing so is especially problematic when one wants to investigate how people come to hold particular views of themselves. Categorization of persons on the basis of an externally observable difference underplays how subjects make themselves and neglects the effects that those persons' actions might have on their senses of themselves. Using social identities as the basis for analysis may be useful as a first step, a sort of gross attempt to make sense of a bewildering array of beliefs and actions that people always hold (e.g., see the analysis in chapter 5). But to end the analysis there is to fail to attend to the many different ways in which people constitute themselves.

Focusing attention on specific social practices relevant to subject formation along a given dimension or facet of identity creates the space for learning more about how actions are connected to beliefs and subject formation. Undoubtedly, practices are always undertaken in the context of institutionalized structures of expectations and obligations, asymmetric political relations, and existing views people have of themselves. But to point to the situatedness of practices and beliefs is not to grant social context a deterministic influence on practice and subjectivity. Rather, it is to ground the relationship among structure, practice, and subjectivity on evidence and investigative possibilities.

Variations in Environmental Subjectivities in Kumaon

I consider two forms of variations in the nature of environmental subjectivity in Kumaon—those that have unfolded over time and those that are geographically distributed across villages. The first set of changes is the puzzle that began this book. From being persons who opposed efforts to protect the forested environment, Kumaonis became persons who undertook the task of protection upon themselves. Instead of protesting the governmentalization of nature, Kumaonis became active partners in it. I describe below the alchemical shift in interests, beliefs, and actions for which the move toward the community partially stands. But it is just as important to understand the differences in relation to environmental practices and beliefs among Kumaonis today and how they affect the costs of environmental regulation.

My examination of changes over time and social-spatial variations in how Kumaon's residents see themselves and their forests draws on three bodies of evidence. The first set of observations is drawn from archival materials about Kumaonis' actions in forests in the first three decades of the twentieth century and then from a survey of forest council headmen in the early 1990s—sixty years after the Forest Council Rules became the basis for local practices.

The second type of evidence comes from two rounds of interviews I conducted with thirty-five Kumaon residents in seven villages. The first round of interviews was conducted in 1989 and the second in 1993.[9] Of the seven villages, four had formed councils between 1989 and 1993. In the interviews, I collected information on villagers' social and economic backgrounds, their practices in relation to their forests, and their views about forests and the environment.

Finally, I use evidence from 244 interviews that I conducted in 1993 in forty-six villages. Of these villages, thirty-eight have forest councils. The councils have adopted various monitoring and enforcement practices. In the remaining eight villages, there are no forest councils, and villagers' relationship to environmental enforcement is restricted to infrequent interactions with forest department guards, who are seldom seen in the forests that these villagers use. Villagers prefer not to see a forest department guard, but even more do they prefer that the guard not see them!

Historical Changes in Environmental Subjectivities

Hukam Singh's personal example illustrates what has obviously been a much larger and more comprehensive process of social environmental change in Kumaon. Recall from chapter 1 the recalcitrant acts of rebellion that hill men performed at the beginning of the twentieth century. Unwilling to accede to the demands of the colonial forest department, often because they were unable to do so, Kumaonis ignored new rules that limited their activities in the forests that the state claimed as its own. They also protested more actively. They grazed their animals, cut trees, and obstreperously set fires in forests that had been classified as reserved. Forest department officials found it next to impossible to enforce the restrictive rules in the areas they had tried to turn into forests.

Law enforcement was especially difficult because of the unwillingness of villagers to cooperate with government officials. The departmental staff was small, the area it sought to police was vast, and the supervisory burden was immense. Decrying lopping for fodder and the difficulty of apprehending those who did it, E. C. Allen, the deputy commissioner of Garhwal, wrote to the commissioner of Kumaon that loppings "are seldom detected at once and the offenders are still more seldom caught red-handed, the patrol with his present enormous beat being probably 10 miles away at the time. . . . It is very difficult to bring an offence, perhaps discovered a week or more after its occurrence, home to any particular village much less individual" (1904: 9). Demarcation of the forest boundaries, prevention of fires, and implementation of working plans meant an impossibly large workload for forest department guards and employees even in the absence of villager protests. When the number of protests was high and villagers set fires often, the normal tasks of foresters could become impossible to perform. One forest department official was told by the deputy commissioner of Kumaon that "the present intensive management of the forest department cannot continue without importation into Kumaon of regular police" (Turner 1924).

After stricter controls were imposed in 1893, the settlement officer, J. E. Goudge, wrote about how difficult it was to detect offenders in instances of firing.

In the vast area of forests under protection by the district authorities the difficulty of preventing fires and of punishing offenders who wilfully fire for grazing is due to the expense of any system of fire protection. Where forests are unprotected by firelines, and there is no special patrol agency during the dangerous season, it is next to impossible to find out who the offenders are and to determine whether the fire is caused by negligence, accident, or intention. . . . The difficulty of making villagers collectively responsible for all fires occurring within their limits in unsurmountable, because the same belt of forest will touch a large number of villages, and how could we make them all responsible in any effective manner? (1901: 10)

In a similar vein, the forest administration report of the United Provinces in 1923 described a fire in the valley of the Pindar River: "During the year, the inhabitants of the Pindar valley showed their appreciation of the leniency granted by Government after the 1921 fire outbreak when a number of fire cases were dropped, by burning some of the fire protected areas which had escaped in 1921. . . . These fires are known to be due to direct incitement by the non-cooperating fraternity" (Anonymous 1924: 266). The sarcasm is clumsily wielded, but its import is obvious all the same. Villagers could not be trusted because ungratefulness was their response to leniency. Other annual reports of the forest department from around this period provide similar claims about the lack of cooperation from villagers, their irresponsibility, and the inadvisability of any attempt to cooperate with them over protectionist goals. At the same time, some government officials underlined the importance of cooperation. Percy Wyndham, asked to assess the impact of forest settlements, said in 1916 that "it must be remembered that in the tracts administration is largely dependent on the goodwill of the people and the personal influence of the officials [on the people]" (quoted in Baumann 1995: 84).

Other reports reveal continuing difficulties in apprehending those who broke government rules. Names of people who set fires could not be identified. Even more unfortunately from the forest department's point of view, not only the common hill residents but also the heads of villages were unreliable. Village heads, known as *padhans*, were paid by the government and were often expected to carry out the work of revenue collection. Their defiance, therefore, was even more a cause for alarm. As early as 1904, the deputy commissioner of Almora, C. A. Sherring, remarked on the heavy workload the patwaris carried for the forest department and argued for increasing their number substantially, as the padhans were unreliable. He wrote to the commissioner: "It is certain that very little assistance can be expected from the padhans, who are in my experience only too often the leaders of the village in the commission of offences and in the shielding of offenders. . . . If the control of open civil forests is to be anything more than nominal we really must have the full complement of patwaris. . . . A large forest staff of foresters and guards is also required" (1904: 2). H. Jackson, the conservator of forests, concurred in a letter to the chief secretary of the United Provinces. According to him: "The headmen of the villages differ little, if at all, from the rest of the population. . . . [There is] no instance of any influential man giving the information regarding forest offences committed in the neighboring forest by inhabitants of his own village, while, on the other hand, it frequently happens that offenders are shielded by *lambardars* (leaders) at the time of an inquiry" (1911: 113). The deputy conservator of forests similarly complained that villagers refused to reveal the culprit in investigations concerning forest-related offenses: "It is far too common an occurrence for wholesale damage to be done by some particular village . . . often nothing approaching the proof required for conviction can be obtained . . . [and] there is too much of this popular form of wanton destruction, the whole village subsequently combining to screen the offenders" (Burke 1911: 44, quoted in Shrivastava 1996: 185).

These reports and complaints by colonial officials in Kumaon make clear the enormous difficulties the forest department faced in realizing its ambition to control villagers' actions on land made into forest. The collective actions of villagers in setting fires and lopping trees and their unwillingness to inform against members of their "fraternity" are clear indications of strands of solidarity that connected them in their work against the colonial state. With unreliable villagers, limited resources, and few trained staff members, it is not surprising that the forest department found it hard to rely only on those processes of forest making that it had initiated and implemented in other parts of India—processes that relied mainly on exclusion of people, demarcation of landscapes, creation of new restrictions, and fines and imprisonment.[10]

The response of the state, in the shape of an agreement with Kumaon residents to create community-managed forests, was the uneasy collaboration among revenue officials, foresters, and villagers that I traced in chapters 3 through 5. This form of regulatory rule has also gone to-

TABLE 8.

Complaints by Forest Council Headmen in Kumaon, 1993 (N = 324)

Complaints Mentioned by Headmen	Number of Headmen Listing the Complaint
1. Inadequate support from forest and revenue department officials	203 (.63)
2. Limited powers of council officials for environmental enforcement	185 (.57)
3. Insufficient resources in forests for the needs of village residents	141 (.44)
4. Low income of the council	130 (.40)
5. Inadequate demarcation of council-governed forests	61 (.19)
6. Lack of respect for the authority of the council	42 (.13)
7. Land encroachment on council managed forests	36 (.11)
8. Lack of remuneration for headmen	31 (.10)
9. Other (e.g., forest boundaries incorrectly mapped, court cases take a long time, residents from other villages break council rules, too much interference in the day-to-day working of the council, lack of information about forest council rules)	64 (.20)

Source: Council Headmen survey by author, 1993.

Note: Figures in parentheses indicate the proportion of headmen mentioning the complaint. Each headman could list up to three complaints.

gether with shifts in how Kumaon's villagers regard forests, trees, and the environment. Colonial forest administrators of the 1900s would have found many of the present-day Kumaon villagers far more interested in forest protection than their counterparts in the early twentieth century. Some indication of the extent to which contemporary Kumaonis have changed in their beliefs, not just their actions, about forest regulation is evident from the results of a survey of forest council headmen I conducted in the early 1990s.

Chapter 1 described the context of that survey. Table 8 summarizes its results. The council headmen in Kumaon have come to occupy an intermediate place in the regulatory apparatus for the environment. On the one hand, they are the instruments of environment-related regulatory authority. On the other, they represent villagers' interests in the forests.

The aggregate numbers in the survey underscore the inferences in chapter 1. The greatest proportion of responses from the headmen concerns the inadequate enforcement support they get from forest and revenue department officials. The government of forests at the level of the community is hampered by the unwillingness or inability of state officials to buttress attempts by villagers to prevent rule infractions. A rough calculation shows that nearly two-thirds of the total responses are directly related to headmen's concerns about the importance and difficulties of enforcing regulatory rules (rows 1, 2, 5, 6, 7, and part of 9). Admittedly, the council headmen are the persons most likely to be concerned about forests and the environment among all the residents of Kumaon. But the point to note is that even when they were presented with an opportunity to freely voice the problems that they face and potential ways of addressing them, only a very small proportion of the responses from the headmen, less than 4 percent, are complaints about their low levels of remuneration (row 8). The headmen evidently put their own interests aside as they tried to grapple with the question of what problems characterized government by communities.

The figures in the table are no more than an abstract, numerical summation of many specific statements that the survey elicited. The plethora of these statements calls for a tabular representation. But the sentiments behind the numbers in the table come from actual words. "I have tried to give up being the head of our committee so many times. But even those who don't agree with me don't want me to leave," observed one of the headmen. Another said: "I have given years of my life patrolling the forest. Yes. There were days when my own fields had a ripening crop [and needed a watchman]. I am losing my eyesight from straining to look in the dark of the jungle. And my knees can no longer support my steps as I walk in the forest. But I keep going because I worry that the forest will no longer survive if I retire." One's own life in exchange for the life of the forest! Sukh Mohan's views about the making and maintaining of his village's forests are centered around his personal contribution, but his commitment to forest protection also matches objectives that the forest department started pursuing more than 150 years ago. Puran Ram gave a reason for his conservationist practices: "We suffered a lot from not having too many trees in our forest. Our women didn't have even enough wood to cook. But since we banned cattle and goats from the forest, it has come back. Now we don't even have to keep a full-time guard. Villagers are becoming more aware."

Puran Ram and Hukam Singh both expressed a hope for a connection between their efforts to conserve the forest and the actions of other villagers. This common hope that I encountered in other conversations as well is an important indication of the relationship between actions and beliefs. It signals that in many of the villages a new form of government frames and enacts reasonable guidelines for villagers' practices in the expectation that over time practice will lead to new beliefs. Villagers subject to regulations crafted and enacted by the councils will come to see that stinting is in their own interest. The forest belongs to the collective defined as the village, and when an individual harvests resources illegally the action adversely affects all members of the collective. The examples of both Puran Ram and Hukam Singh, as, indeed those of more than two-thirds of the headmen in my survey, suggest that the expectation is not just a fantasy.

The differences in the voice and tenors of the archival and more recent statements I collected offer a basis for making the judgment that the practices and views of many of Kumaon's residents about their forests have changed substantially. These changes occurred after the passage of the Forest Council Rules in 1931. Partly responsible for these changes is the idea that Kumaonis can consider the region's forests their own once again.[11] But within this shift in ownership by the collective there are many variations in beliefs. These variations in beliefs about forests and practices related to forest protection are not systematically connected to the benefits people receive. Benefits from forests are formally allocated equitably, and this equitable allocation is reflected in the harvests of most villagers. But even within villages there is significant variation in how villagers see forests and/or try to protect them.

It may be argued that appropriations by the colonial state in the early twentieth century drove a wedge between forests and villagers. Subsequently, the rules that led to community-owned and managed forests reaffirmed the propriety and legality of villagers' possession of them. They recognized that villagers have a stake in what happens to forests and expressed some faith in their ability to take reasonable measures for protection, especially with proper guidance. These institutional changes go together with changes in villagers' actions and beliefs about forests. One way to explain this change is to suggest that the shift in policy and subsequent changes in beliefs and actions are unrelated.

They are sufficiently separated in time that a causal connection can only tenuously be drawn. This is frankly unsatisfactory. At best, it is a strategy of denial. It is also incapable of explaining the evidence that follows in the next section. A more careful argument would at least suggest that shifts in villagers' actions and statements in the latter part of the twentieth century are no more than a response to changes in interests that the new policy naturally generated. The transfer of large areas of land to villagers in the form of community forests has created in them a greater concern to protect the forests and care for the vegetation they control.

This is an important part of the explanation. It usefully suggests that the way social groups perceive their interests is significantly dependent on policy and government instead of being constant and immutable. But it is still inadequate in two ways. It collapses the distinction between the interests of a group as perceived by an observer-analyst, and the actions and beliefs of members of that group. In this explanation, interests, actions, and beliefs are of a piece, and any changes in them take place all at once. This assertion of an identity among various aspects of what makes a subject and the simultaneity of change in all of these aspects are at best difficult propositions to swallow. The difficulty can in part be illustrated by one's own experience. We often arrive at a new sense of what is in our interests but continue to hold contradictory beliefs and act in ways that match more closely the historical sense of our interests. Many of the headmen I interviewed in Kumaon, or who were a part of my survey, tried to enforce rules that they knew were not in the interests of their own households. Their wives and children were often apprehended by the forest guards they had appointed. Yet they defended their actions in the name of the collective need to protect forests and expressed the hope that over time villagers would come around to their view and change their practices. As the next section makes clear, their hopes were not in vain. Many villagers proved susceptible to these shifting strategies of government.

A second problem with the explanation that headmen care for forests because they have the right to manage them is that it confuses the private interests and actions of the headmen with their public office and interests. The forests that have been transferred to village communities are managed by collective bodies of anywhere between 20 to 200 villagers, who are represented by the forest councils and their headmen

(Agrawal 2000). To attribute a collective interest to these bodies and to explain what the headmen of these councils say in terms of that interest is to elide all distinctions between specific, individual actors and the organizations they lead. A more careful exploration of other actors in Kumaon who are involved with the local use and protection of forests is necessary. Only then can we begin to make sense of the changes indicated by the survey of headmen summarized in table 8 and the information provided below about beliefs Kumaonis hold about their forested environment.

Recent Changes in Environmental Subjectivities

When I went to Kumaon and Garhwal in 1989, I traveled there as a student interested in environmental institutions and their effects on the actions and beliefs of their members. My main interest was to show that environmental institutions—the forest councils—had a significant mediating effect on the condition of forests (Agrawal and Yadama 1997). Not all villages had created local institutions to govern their forests. Of the thirteen that I visited, only six had forest councils. The ones that did varied in the means they used to protect and guard forests. Since my interest was primarily to understand institutional effects on forests, I focused on gathering archival data from records created and maintained locally by village councils (Agrawal 1994).

My conversations with village residents were chiefly intended to gain a sense of their views about forests and the benefits they provide to residents. I found that villagers who had forest councils were typically more interested in protection. They tried to defend their forests against harvesting pressures from other residents within the same village but especially from those who did not live in their village. They also stated clear justifications of the need to protect forests, even if their efforts were not always successful. In one village near the border between Almora and Nainital districts, a villager used the heavy monsoons to make the point: "Do you see this rain? Do you see the crops in the fields? The rain can destroy the standing crop. But even if the weather were good, thieves can destroy the crop if there are no guards. It is the same with the forest. You plant a shrub, you give it water, you take care of it. But if you don't protect it, cattle can eat it. The forest is for us, but we have to take care of it if we want it to be there for us."[12] Another villager pointed to the difficulties of enforcement in a council meeting I attended.

> Until we get maps, legal recognition, and marked boundaries [of the local forest], council cannot work properly. People from Dhar [a neighboring village] tell us that the forest is theirs. We cannot enter it. So we can guard part of the forest, and we don't know which part [to guard]. Since 1984, when the panchayat was formed, we have been requesting the papers that show the proper limits so we can manage properly, protect our forest. But what can one do if the government does not even provide us with the necessary papers?[13]

A second villager in the same meeting added: "Mister, this is Kaljug.[14] No one listens to the *sarpanch* [council headman]. So we must get support from the forest officers and revenue officers to make sure that no one just chops down whatever he wants."

On the other hand, residents of the seven villages that did not have forest councils scarcely attempted any environmental regulation—no doubt because the forests around their villages were owned and managed by either the forest department or the revenue department. Villagers perceived regulation as the responsibility of the state and a constraint on their actions in the forest—gathering firewood, grazing animals, harvesting trees and nontimber forest produce, and collecting fodder. There were thus clear differences in the actions and statements of villagers who had created forest councils and brought local forests under their control and villagers who relied on government forests to satisfy their requirements for fodder and firewood.

During my return visits in 1993, I realized that four of the seven villages (Pokhri, Tangnua, Toli, and Nanauli) without forest councils in 1989 had formed councils in the intervening years. They had drafted constitutions modeled on others in the region. Under the provisions of the Forest Council Rules, they had brought under their control the local forests that had been managed by the departments of revenue or forests. A series of resolutions created new rules, which became the basis for governing forests. These resolutions prescribed how (and how often) to hold meetings, elect new officials, allocate fodder and grazing benefits, set levels of payments by villagers in exchange for the right to use forests, monitor forest conditions and use, and sanction rule break-

TABLE 9.

Environmental Beliefs of Villagers in Kumaon in 1989 (N = 35)

	Agreement with the Statement "Forests Should be Protected" (1 = low, 5 = high)	Reason to Protect the Forest (economic or environmental)	Willingness to Reduce Family Consumption of Forest Products (1 = low, 5 = high)
Residents of villages that had created forest councils (Pokhri, Tangnua, Toli, Nanauli) (N = 20)	2.35	Economic = 16 Environmental = 4	1.45
Residents of villages that did not have a forest council (Darman, Gogta, Barora) (N = 15)	2.47	Economic = 11 Environmental = 4	1.73

Source: Field work, 1989.

Note: Respondents were surveyed in the villages of Pokhri, Tangnua, Toli, Nanauli, Darman, Gogta, and Barora.

ers. Exposure to these new institutional constraints, council members hoped, would lead villagers to adopt more conservationist practices.

In these four villages, I had talked with twenty residents in 1989. At that time, their statements did not suggest that they felt any pressing need to conserve the environment. Little distinguished their actions and views from those of the fifteen residents with whom I had talked in the other three villages (Darman, Gogta, and Barora). I had asked my thirty-five respondents three questions about their views on the environment. Table 9 reports the responses of villagers to the three questions in summary form.

I. Do you agree with the statement, "Forests should be protected." Please indicate the extent of your agreement by using any number between 1 and 5 where 1 indicates a low degree of agreement, and 5 indicates strong agreement.

II. If forests are to be protected, should they be protected for 1) economic reasons such as their contribution of fodder, firewood, and green manure, or for 2) other non-economic benefits they provide including cleaner air, soil conservation, and water retention.

III. Do you agree with the statement, "To protect forests, my family and I are willing to reduce our consumption of resources from the local forests." Please indicate the extent of your agreement by using any number between 1 and 5 where 1 indicates a low degree of agreement, and 5 indicates strong agreement.

The first question tries to elicit villagers' responses to a very general statement about forest protection. The second examines the extent to which villagers see forests as an environmental versus a primarily economic resource. And the third question inquires into the willingness of villagers to endure some constraints so as to meet the objective of forest conservation. The figures in the table indicate that the differences among the residents of the seven villages are relatively minor. All villagers expressed limited agreement with the idea that forests should be protected. Their reasons were mainly economic. And they were relatively unwilling to place any constraints on the consumption of their families to ensure forest conservation.

Although there is little basis for differentiating the responses of both sets of villagers in 1989, changes became evident in 1993 when I talked to the same villagers again. In the case of residents of the four villages that had created forest councils, the differences were obvious both in their actions and in what they said about forests and the environment. Some had come to participate actively in their new forest councils. They attended council meetings. A few had limited their use of the village forest. Some acted as guards. They even reported on neighbors who had broken the council's rules. The similarities in their changed behavior and the changed behavior of forest council headmen briefly described above (and also in chapter 1) are quite striking. Those who had come to have a forest council in their villages, or, perhaps more accurately, those whose councils had come to have them, had begun to view actions in forests differently.

Of course, there were others in these four villages whose views had not changed so much. Those with whom I talked were especially likely to continue to say and do similar things as in 1989 if they had not participated in any way in the formation of the forest councils or in the suite of strategies councils used to try to protect forests. If they had

TABLE 10.

Changing Beliefs of Kumaon's Villagers about the Environment, 1989–93 (N = 35)

	Agreement with the Statement "Forests Should be Protected" (1 = low, 5 = high)	Reason to Protect the Forest (economic or environmental)	Willingness to Reduce Family Consumption of Forest Products (1 = low, 5 = high)
Residents of villages with forest councils in 1989 (Pokhri, Tangnua, Toli, and Nanauli)	2.35	Economic = 16 Environmental = 4	1.45
Residents of Pokhri, Tangnua, Toli, and Nanauli in 1993	3.65	Economic = 12 Environmental = 8	3.00
Residents of villages without forest councils in 1989 (Darman, Gogta, and Barora)	2.47	Economic = 11 Environmental = 4	1.73
Residents of Darman, Gogta, and Barora in 1993	2.27	Economic = 12 Environmental = 3	1.87

Source: Field research, 1989, 1993.

Note: Respondents were surveyed in the villages of Pokhri, Tangnua, Toli, Nanauli, Darman, Gogta, and Barora.

become involved in the efforts to create a council or protect the forest that came to be managed by the council, they were far more likely to suggest that the forest requires protection. They were also more likely to say that they were willing to become personally invested in protection. This is certainly not to claim that participation in council activities is a magic bullet that necessarily leads to transformation of subject positions. And yet the testimony of these twenty residents, by no means a representative sample in a statistical sense, constitutes a valuable window on the changing beliefs of those who come to be involved in practices of environmental regulation (see table 10).

The first two rows of the table make it clear that on the average residents in the four villages expressed greater agreement with the idea of forest protection and a greater willingness to reduce their own consumption of products from local forests than they had in 1989. Of the twenty individuals, thirteen had participated in monitoring or enforcement of forest council rules in some form, and the shifts in their environmental beliefs turned out to be stronger than those of villagers who were not involved in any council-initiated actions.

The information from interviews in these four villages is especially useful compared to the fifteen interviews in the three villages where no councils had emerged in the intervening years. In these three villages, where I also conducted a second round of interviews in 1993, there had been little change in the environment-related practices and views of local residents. They still regarded, and often rightly so, the presence of government guards to be a veritable curse. Many of them, usually after looking around to make sure no officials were present, roundly abused the forest department. Indeed, this is a practice that villagers in other parts of rural India may also find a terrifying pleasure. But even when my interviewees agreed that it was necessary to protect forests because of the many benefits trees provide, they were unwilling to do anything themselves to further such a goal. For the most part, their beliefs about forests and the environment had also changed little. The last two rows of table 10 suggest that there have been only very minor shifts in villagers' perceptions about the need to protect forests and in their willingness to work toward that goal at some cost to themselves.

Variations in Environmental Subjectivities: The Place of Regulation

The practices and perceptions associated with the emergence of forest councils in Kumaon contain many variations. The preceding discussion contains some important clues as to the sources of these variations. But the information in tables 9 and 10 is highly aggregated and does not have sufficient texture to provide much insight into how practices related to environmental regulations affect the way villagers think about their actions in forests. Further, it is not just the historical dimension of the production of different forms of environmental subjects that needs attention; it is also important to explore the contemporary differences in the making of environmental subjects.

To examine how and to what extent regulatory practices relate to the environmental imaginations of Kumaonis, I report on the responses of more than two hundred persons I met and interviewed in 1993. Since the number of people I interviewed was much larger in 1993 than in 1989, it is possible to examine how different forms of monitoring and enforcement relate to my respondents' beliefs about the environment. Recall from the previous chapter that the forest councils in Kumaon use five different forms of monitoring and enforcement in their forests (see figure 7). Two of these fall under the heading of mutual monitoring. In one, each village resident can monitor all the others and report illegal actions in the forest to the council. In the second, households are assigned monitoring duties in turn. Under mutual monitoring, there is little specialization in the task of monitoring and none of the monitors is paid for the services he performs. Three forms of third-party monitoring exist, and in all of them a specialized monitor is appointed for specific durations and is paid for the work performed. Forms of third-party monitoring are distinguished by the mode of payment to the guard: direct payments by households in cash or kind, salary payments by the council but from funds raised locally, and salary payments from funds made available through commercial sales of forest products or transfers received from the state. Table 11 summarizes the responses for different forms of monitoring and shows the extent to which participation in monitoring and enforcement is connected to respondents' beliefs about forests and the environment.

Table 11 provides information on the same three questions as did tables 9 and 10, but it probes deeper. Instead of simply reporting the average responses of all interviewed individuals, the table presents the answers of respondents by separating them according to the monitoring regulations adopted by forest councils and their involvement in monitoring practices. Thus, the table relates the participation in particular forms of environmental enforcement to villagers' reported beliefs about the environment. To interpret the table, consider row 1. It provides information on a total of ten respondents, only eight of whom participated in mutual monitoring (all villagers monitor all other villagers). The answers of these respondents to the three questions listed earlier are arranged according to whether they participated or abstained in mutual monitoring. Rows 1–5 follow this pattern. The last row is for villages that did not have their own council-governed forest, so their residents had no opportunity to participate in environmental monitoring.

TABLE 11.

Participatory Practices and Environmental Beliefs of Kumaon's Villagers, 1993
(N = 244)

Monitoring Strategy Used by the Forest Council[a]	Number of Villagers[b]	Agreement with the Statement "Forests Should be Protected" (1 = low, 5 = high)	Reason to Protect the Forest (economic or environmental)	Willingness to Reduce Family Con-sumption of Forest Products (1 = low, 5 = high)
MUTUAL MONITORING				
All households monitor all the time (2)	Total = 10			
	Participant (8)	3.25	Eco = 4 Env = 4	2.63
	Nonparticipant (2)	3.00	Eco = 2 Env = 0	2.00
Households are assigned in rotation (3)	Total = 17			
	Participant (12)	4.25	Eco = 4 Env = 8	3.42
	Nonparticipant (5)	2.80	Eco = 4 Env = 1	2.40
THIRD-PARTY MONITORING				
Households make direct payments to monitor (7)	Total = 39			
	Participant (32)	4.00	Eco = 20 Env = 12	3.06
	Nonparticipant (7)	2.86	Eco = 6 Env = 1	2.29
Monitor's salary is paid from locally raised funds (18)	Total = 98			
	Participant (55)	3.98	Eco = 36 Env = 19	2.80
	Nonparticipant (43)	2.81	Eco = 38 Env = 5	1.72
Monitor's salary is paid from external funds (8)	Total = 41			
	Participant (9)	3.66	Eco = 6 Env = 3	2.66
	Nonparticipant (32)	2.31	Eco = 30 Env = 2	1.53
No forest council, no monitoring (8)	Total = 39	2.33	Eco = 31 Env = 8	1.74

Source: Field research, 1993.

[a] Figures in parentheses indicate the number of councils.

[b] Figures in parentheses indicate the number of respondents.

It is striking that for all forms of monitoring respondents expressed a greater need to protect forests if they participated in monitoring than if they did not. But the difference between participants and nonparticipants is more striking as enforcement and monitoring become more specialized and when villagers participate directly in enforcement. So when we move to rows 2, 3, and 4, where monitoring is a specialized role for assigned households or for an assigned individual who acts as guard and whom villagers pay out of their own funds, participation in monitoring is positively related to both a greater appreciation of the need to protect the environment and a greater willingness to undergo some limits on personal consumption to protect the environment. The table thus suggests that the difference between those who participate in monitoring and those who do not is greatest in those forms of monitoring where there is role specialization and villagers directly invest labor or money in monitoring.

This inference is in line with the expectation that villagers' beliefs about the environment are likely to be in accord with their practices. The table shows that the choice of a monitoring method by a forest council does not affect all villagers in the same manner. It is those villagers who take a direct role in monitoring activities or in funding monitoring who express the greatest interest in forest protection. These villagers are also far more invested than nonparticipants in seeing forest protection as an important goal, even if no economic benefits are expected. The responses of nonparticipants in each type of monitoring are close to those of villagers who do not have a forest council in their village at all.

Further, it is in villages where there is the greatest participation in monitoring and enforcement that councils have the greatest ability to raise resources to protect forests. Both in those villages where the most basic form of mutual monitoring is in force and in villages where resources for monitoring are primarily secured from outside sources, the ability of the council to gain villagers' participation is limited. As table 11 shows, forest councils represented in row 5 secure the lowest levels of participation from their villagers. It is also in villages represented in rows 2, 3, and 4 that residents express the greatest desire to protect forests and are willing to undergo some personal hardship to accomplish forest protection.

Note that I am not using the evidence in table 11 to theorize a causal-sequential relationship between participation in monitoring and the development of environmental subjectivities. Such an inference would only be possible by interviewing the same respondents before and after their participation in enforcement. The information in table 10 comes closest to such before and after evidence. The figures in table 11 only show variations in subjectivities across different forms of monitoring and participation in monitoring. It would be reasonable to suggest that it is differences in beliefs that prompt my respondents to participate in monitoring rather than participation that leads them to different beliefs. It is when we consider together the evidence in tables 10 and 11 that it becomes at all justifiable to suggest that there are variations in the environmental sensibilities of Kumaon residents that are systematically related to their participation in environmental enforcement and that these differences stem at least to some extent from such participation. But in any case even table 11 shows important differences in how villagers in Kumaon think about their forests and their relationship with the environment.

The importance of participation in different monitoring mechanisms becomes evident also in comparison with social identity categories such as gender and caste. Consider the information in table 12. It shows the difference between environment-related beliefs of interviewed villagers by their gender (women versus men), caste (high versus low), and participation in different forms of monitoring. There is scarcely any difference between men and women or respondents of higher or lower castes. They seem equally (un)likely to want to protect forests or reduce their household's consumption to conserve forests. On the other hand, those who are involved in some form of monitoring and environmental enforcement are more likely to agree with the need to protect forests, to say that forests need to be protected for environmental rather than economic reasons, and to accept some reduction in their own use so as to ensure forest protection.

It is reasonable to conclude that when villagers participate in monitoring and enforcement they come to realize at a personal level the social costs generated by those who do not adhere to the practices and expectations that have been collectively chosen. They confront more directly those who act illegally in the forest and then must decide whether to ignore such confrontations, to choose more carefully to enforce what they had agreed to do collectively, or to join others who

TABLE 12.

Gender, Caste, and Participation and Their Relationships with
Beliefs Concerning the Environment, 1993 (N = 205)

Monitoring Strategy Used by the Forest Council, Dimension of Difference[a]	Agreement with the Statement "Forests should be protected" (1 = low, 5 = high)	Reason to Protect Forest[b]	Willingness to Reduce Family Consumption of Forest Products (1 = low, 5 = high)
GENDER			
Women (95)	3.38	Economic = 69 (73)	2.45
		Environmental = 26 (27)	
Men (110)	3.36	Economic = 80 (72)	2.34
		Environmental = 30 (28)	
CASTE			
High (106)	3.44	Economic = 78 (74)	2.44
		Environmental = 28 (26)	
Low (99)	3.30	Economic = 71 (71)	2.42
		Environmental = 28 (29)	
PARTICIPATION			
Yes (116)	3.92	Economic = 70 (60)	2.97
		Environmental = 46 (40)	
No (89)	2.66	Economic = 79 (89)	1.74
		Environmental = 10 (11)	

Source: Field research, 1993.
[a]Figures in parentheses indicate number of respondents.
[b]Figures in parentheses indicate percentage of respondents.

are violating socially constructed norms and expectations. Opting for the first two options is also to opt for what I am nominating an environmental subjectivity. Similarly, those who in their actions violate collectively generated guidelines to regulate practice can often continue to do so when it is individually expedient and there is no mechanism to regulate them. But when enforcement is commonplace they are often confronted with the knowledge of their own deviations from what they had agreed to do. When their actions are met with direct challenges that they count as appropriately advanced, it becomes far more difficult to continue to act and believe in a divergent manner. It is in examining

TABLE 13.

Contributions per Household toward Enforcement by Forest Councils, 1993
(N = 205)

Form of Monitoring	Number of Respondents	Contributions per Household (in rupees)
Mutual monitoring (each household monitors all others)	10 (2 villages)	9.33
Mutual monitoring (households assigned monitoring duty in rotation)	17 (3 villages)	11.44
Third-party monitoring (households pay monitors directly)	39 (7 villages)	36.61
Third-party monitoring (salary paid out of locally raised funds)	98 (18 villages)	19.98
Third-party monitoring (salary paid out of external transfers)	41 (8 villages)	16.22

Source: Forest council records (data collected in 1993).

practices of villagers closely that it becomes possible to trace the links
between politics, institutional rules, and subject formation.

The effects of more widespread participation are also visible in the
resources that councils are able to mobilize for protecting forests. Consider table 13. It presents the per household contributions that forest
councils are able to use in a year.

The form of monitoring that leads to the highest levels of contributions is one where villagers pay the guard directly. Mutual monitoring
produces the lowest levels of contributions. This should not surprise us
because councils resorting to this form of forest protection have been
unable to gain the agreement of their members to expend any monetary or material resources on monitoring. The "contributions" mentioned under option 5 (third-party monitoring, in which the guard is
paid from external funds) are misleading because these are, strictly
speaking, the resources available for monitoring from all sources, not
just the contributions from village households.

Clearly, engagement with the regulatory practices of monitoring and
enforcement is positively connected both to the existence of environmental orientations among Kumaon's residents and to higher mone-

tary and material contributions toward enforcement per household. The most important inference for policy is that certain forms of environmental enforcement are simultaneously associated with higher levels of involvement by villagers and the generation of environmental subjectivities.

Intimate Government

> Power had to be able to gain access to the bodies of individuals, to their acts, attitudes, and modes of everyday behavior. . . . But at the same time, these new techniques of power needed to grapple with the phenomena of population, in short to undertake the administration, control, and direction of the accumulation of men.
> —Michel Foucault, *Power/Knowledge*

A useful metaphor that helps to think about the mechanisms that underpin the production of various forms of subjectivity in Kumaon is what Latour (1987) has called "action at a distance" and, following him, Miller and Rose have termed "government at a distance" (1990). Latour shows how intentional causes could operate at a distance to effect particular kinds of actions in places and by people who are not directly controlled. Examining the work of scientists, Callon and Latour (1981) and Latour (1986) describe the affiliations and networks that help establish links between calculations at one place and actions in another. The crucial element in their argument is the "construction of allied interests through persuasion, intrigue, calculation, or rhetoric" (Miller and Rose 1990: 10). It is not that any one of the actors involved appeals to existing common interests. Rather, one set of actors, by deploying a combination of resources, convinces another that the goals and problems of the two are linked and can be addressed using joint strategies.[15]

In Kumaon, two crucial types of resources that the forest and revenue departments combined and deployed in the 1920s and 1930s were information and forests. Information about the adverse effects of central government of forests in Kumaon during the second decade of the twentieth century and about existing government of forests by communities in the region prepared the ground for the argument that regulatory control over forests could be decentralized to positive effect. The experience of decentralized government of forests in Burma and Madras and the investigation of these experiences firsthand by departmental officials in the 1920s helped produce the design of the Forest Council Rules of 1931. The gradual return of the same forested lands that villagers had used until the 1890s (those that the Kumaon forest department appropriated between 1893 and 1916) provided the material basis for the idea of a common interest in forest protection between village communities in Kumaon and the forest department. Forest councils became the institutional means of pursuing this common interest over long geographical distances.

In the formulation "action at a distance" or "government at a distance," it is geographical distance that action and government overcome. In an important sense, these formulations are about the uncoupling of geographical from social and political distance that forms of modern government accomplish. By clarifying and specifying the relationship between particular practices in forested spaces and the sanctions that would follow those particular practices, government encourages new kinds of action among those who are to be governed. Action at a distance thus overcomes the effects of physical separation by creating regulations known to those located at a distance. Officials who oversee the translation of these regulations onto a social ground succeed in their charge because of the presence of a desire among environmental subjects to follow new pathways of practice.

One can well argue that government of the environment in Kumaon conformed to this logic of action at a distance in its earlier phases—before the institution of community-based government. However, in this earlier phase the effort to induce a change in the actions of villagers failed because of the inability of government to constitute a vision of joint interests in forests with which Kumaonis could identify. The forest councils created the potential for villagers and state officials to come together in a new form of government through which a compelling vision of joint interests could be manufactured. Not all villagers created forest councils, however, and even in villages where forest councils came into being not all were equally successful. An additional development was necessary to make the government at a distance symbolized by forest councils succeed. Once Kumaon's villagers had crafted highly dispersed centers of environmental authority for and by themselves, processes of government at a distance came to be supplemented simultaneously by what might be called "intimate government."[16]

In contrast to government at a distance, which presupposes centers of calculation, constant surveillance, continuous collection of information, unceasing crunching of numbers, and the imposition of intellectual dominance through expertise (Miller and Rose 1990: 9–10), intimate government in Kumaon works by dispersing rule and scattering involvement in government more widely. In consequence, there are numerous locations of decision making in each of which there are actors who work in different ways and to different degrees to protect forests. Homogeneity across these locations is difficult to accomplish for a variety of reasons, among them, differences in levels of migration, histories of cooperation, social stratification, occupational distribution, and resource endowments. Monitoring of villagers' actions is patchy and unpredictable. Councils collect information, but it is available only locally; it is seldom processed and presented in a way that may be useful for policy-making elsewhere. Practice and sociality rather than expertise form the basis of intimate government to regulate villagers' actions. The ability of regulation to make itself felt in the realm of everyday practice is dependent on channeling existing flows of power within village communities toward new ends related to the environment. The joint production of interests is based on multiple, daily interactions within the community. As the community becomes the referential locus of environmental actions, it also becomes the arena where intimate government unfolds.

Intimate government shapes practice and helps to knit together individuals in villages, their leaders, state officials stationed in rural administrative centers, and politicians interested in classifying existing ecological practices. Intimate government is about the creation and deployment of links of influence between a group of decision makers within the village and the common villagers whose practices these decision makers seek to shape. Institutional changes in the exercise of power are the instruments through which these links between decision makers and the practices of villagers are made real. When successful, they are closely tied to processes of environmental enforcement, as the evidence in this chapter suggests. Variations in institutional forms of enforcement are connected to the participation that villagers are willing to provide and forest council decision makers try to elicit. Specialization of enforcement roles and direct participation in enforcement seem to create the greatest willingness on the part of villagers to contribute to environmental enforcement as well as to express an interest in environmental protection, as tables 9 through 12 make clear.

But not all forms of institutional enforcement are equally available to all forest councils, as was discussed in chapter 5. If the number of households in a village is small and the households are relatively poor, the ability of villagers to contribute toward the payment of a salary is limited. If a village is highly stratified or there are many disagreements among the villagers, they are also less able to enact sustainable environmental enforcement. These variations in village-level processes influence the extent to which different village communities are able successfully to take advantage of the state's willingness to disperse rule and decentralize control over forests.

Intimate government is also about the ways in which villagers themselves try to shape their conduct in forests. Government at a distance works in Kumaon only in conjunction with intimate government because villagers get involved in regulatory practices that they see as important to their own long-term interests. With the redefinition of interests that exposure to scarcities and regulations makes explicit, a calculation of the costs and benefits of illegal harvests from their own forests versus those from government or other communities' forests has come to pervade the environmental practices of households. Instead of simply harvesting the fodder, firewood, or the timber they need from forests near their homes, Kumaon's residents now reckon carefully before deciding where, how, how much, and when to harvest. The experience of scarcity makes such reckoning unavoidable.

At the same time, it is not simply constraint that new forms of community-based government embody.[17] If regulations necessitate careful estimations of availability and scarcity, they also go together with possibilities for other kinds of corrective action against decision makers. If villagers do not favor the way their forest is being governed by their councils, they can attempt to change the regulations adopted by their council members or even to change the council's membership. Channels that affect what happens in forests allow influence to flow in multiple directions rather than only one way. And the everyday regulation of what happens in forests is influenced far more directly by villagers' links to officials in their forest councils than by their links to state officials in the forest and revenue departments.

Although intimate government in Kumaon's communities has helped efforts by the state to govern at a distance, it has done so in ways that local residents believe to be defined locally. Villagers may protect forests and control illegal practices of harvesting and extraction. They may also use the language of regulation and many of the same idioms of protection that state officials deploy. But they do so in pursuit of goals that they imagine as their own and in which they often construct state officials as inefficient, unsupportive, or corrupt. This imagined autonomy, stemming from precisely those practices of conservation encouraged by state officials, is crucial to the success of decentralized protection.

My focus on how variations in practices of participation relate to variations in subjectivities tries to move away from abstract, static categories of social classification based on caste, gender, or location. The many variations in the nature of regulatory practices within villages, among men and women, and in upper and lower castes render such classifications partially useful at best. Terms such as *cultural forms* and *symbolic systems*, which are central to Paul Willis's penetrating study of the reproduction of the difference between capitalists and workers, seem equally distant from the process of subject making. Willis is concerned with similar questions about the "construction of subjectivities and the confirmation of identity" (1981: 173). But it is in the examination of the actual practices of schooling among "working class kids" rather than in its abstract cultural-Marxist theoretical structure, that his study finds its most compelling insights.

The previous section, in focusing on the aggregate environmental responses of Kumaon's residents, suggested that using social categories such as gender and caste to try to understand subject formation serves only to *obscure* the processes through which subjects are made. These categories are useful only as proxies, hinting at a small fraction of the interactions that go into the making of environmental subjects. A shift away from categorical relations toward villagers' involvement in practices of social-ecological regulation helps to uncover the frame that holds the three conceptual units of analysis in this book together: politics, institutions, and subjectivities.[18] The focus on practice shows that these seemingly different concepts are linked in the everyday lives of Kumaon's villagers. It is in the investigation of the texture of social practice, simplified analytically by a focus on forms of monitoring and enforcement, that it becomes possible to see how environmental politics is lived by those who are subject to it.

Cultivating Environmental Subjects

The argument that there are relationships between government and subject formation and policy and subjectivity (Foucault 1982: 212) has come to be well rehearsed in the wake of Foucault's original insights (Cruikshank 1994; Hannah 2000; Mitchell 2000; Rose 1999; Tully 1988). This relationship can be traced by examining technologies of power that form subjects and encourage them to define themselves in particular ways and technologies of the self that individuals apply to themselves in order to transform their own conditions (Miller 1993: xiii–xiv). Both of these technologies are united in the idea of government based on knowledge, and they are visible in the processes that unfolded in the making of environmental subjects in Kumaon. My discussion of these processes shows that the relationship between government and subject formation is one of mutuality and dependency rather than agonism and autonomy.

This chapter has chosen not to engage the friction and heat that discussions about Foucault's ethics often generate. Although it is surely important to examine whether his concept of power and the subject leads to an inability to criticize social phenomena, what is more interesting for my purposes is the extent to which some of Foucault's later ideas about government and its relationship to subject formation can be investigated on an evidentiary basis in the context of variations in environmental subjecthood in Kumaon.[19] Foucault is often accused of making provocative conceptual innovations that cannot be deployed in relation to evidence generated from a social ground. Similarly, much political-philosophical debate on subject formation proceeds as if subjects emerge and exist independent of historical, political, and social grounds. It thus constantly runs the risk of becoming irrelevant to actual processes of subject formation. This chapter has attempted to place Foucault's ideas about subject formation against a social and political context and to address subject formation concretely rather than abstractly. Although in the process I have simplified the conceptual architecture of philosophical discussions about the subject, I have done so with a view to focusing carefully on a dilemma that divides much social-theoretical discussion about the subject. More concretely, this chapter has tried to show what differentiates various kinds of subjects by viewing practice as the crucial link between power and imagination. It has examined how close attention to practice can permit the joint examination of seemingly different, abstract constructs such as politics, institutions, and subjectivities.

In this context, Butler's (1997: 10) caution against using the term *subject* interchangeably with *person* or *individual* needs to be taken seriously. Her warning is most useful for the recognition that the relations of power within which subjects are formed are not necessarily the ones they enact upon formation. The temporal sequentiality she introduces in the relationship between subjects and power helps underline the fact that the conditions of origin of a subject need have no more than a tenuous impact on that subject's continuing existence and actions.[20] In Kumaon, the production of environmental subjects in the early twentieth century within the forest department, one might note, led to a cascading series of changes in institutional, political, and social domains connected to the idea of community. It is in this realm of the community that new environmental subjects have emerged.

The making of environmental subjects in Kumaon is thus part of a broader dynamic that new forms of government have unleashed in Kumaon. In tracing the making of environmental subjects, the earlier sections in this chapter examined how variations in subjectivities relate to participatory practices around different forms of environmental monitoring and enforcement. The question of subject formation, implicit in most studies of environmental government, is crucially connected to participation and practice. The practices of enforcement and regulation in which villagers have come to participate are about a more careful government of environment and of their own actions and selves.

Thus, the emergence of environmental subjects in Kumaon's villages has been as much a consequence of processes marked by government at a distance as it has been about intimate government. The state's efforts to govern at a distance made available to villagers the possibility of forming forest councils. The recognition of a mutual interest in forests that was brought into existence by the seeming concessions of the state led some village communities to constitute themselves as creators of forest councils. Simultaneously, the willingness of the forest councils to initiate various processes of intimate government in their own communities has affected their success in gaining villagers' participation and the extent to which village residents would turn into environmental subjects.

7

Conclusion: The Analytics of Environmentality

A steam-engine moves. The question is asked, How is it moved? A peasant answers, It is the devil moving it. Another man says, The steam-engine moves because the wheels are going round. A third maintains that the cause of the motion is to be found in the smoke floated from it by the wind.—Leo Tolstoy, *War and Peace*

The preceding chapters documented and analyzed some remarkable shifts in environmental government, sociality, and subjectivities that have occurred in Kumaon over the last century and a half. I suggested in part 1 of the book that the transformations I have examined are rooted in the emergence of two underlying discursive beliefs about the natural world: (1) nature is an entity discrete from humans and endangered by reckless human actions; and (2) this endangered nature needs protection, which can be generated in the form of careful government.[1] With increasing knowledge of the natural world and the awareness that humans can affect natural processes in an unprecedented fashion, social pressures in favor of protecting nature have grown. But with each additional piece of evidence suggesting the pervasive impact of humans on their world, utopian and romantic visions of an unsullied nature have given way to pragmatic programs for protecting the environment.

I have argued, thus, that in Kumaon the government of nature led to the birth of the environment. Indeed, one of the implicit points that helped frame the discussion in part 1 of the book is that the increasing intensity of care for and government of nature helped the idea of environment to emerge. The use of numbers and statistics helped refine the way the government of nature would work. My discussion has focused on vegetation and forests to examine the strategies through which environmental government remakes nature. But the use of statistics and numbers to shape the making of forests in India and Kumaon, discussed in chapters 2 and 3, occurred together with other far-reaching changes. New technologies of government dispersed centers of environmental regulation and ecological decision making throughout the region and redefined the political relationships between what might broadly be called the state and the local (chapter 4). A more intimate and precise regulatory rule took shape as villagers began to take care of their forests in collaboration with state officials (chapter 5). Kumaon also witnessed the making of new environmental subjects (chapter 6). The emergence of environmental subjects in Kumaon has involved a complex interaction between the way local residents have understood their relationship with forests and the contexts within which their understandings have become possible. Thus, transformations in knowledges, politics, institutions, and subjectivities have been crucial to the character of emerging environmental politics.

The centrality of knowledges, politics, institutions, and subjectivities to changing aspects of environmental politics in Kumaon provides a vantage point from which we can examine recent shifts in the nature of environmental politics elsewhere and attempts by scholars to theorize this field of study. Especially since the early 1980s, the nature of environmental politics has shifted as global geopolitical circumstances have changed and nation-states have come to recognize the limitations of centralized forms of government (Bates 2001; Herbst 2000).

As a result, more than fifty developing countries have moved toward environmental regimes similar in their general outline to the kind of locally based government of forests in Kumaon that this book examines (Agrawal 2001b; FAO 1999). The central feature of these new environmental regimes is a closer involvement of those who depend on various environmental goods—forests, fisheries, pastures, and irrigation waters, among others—in the government of environment. A reorganization of institutional arrangements has facilitated changes in environmental practices and levels of involvement in government.

Such ongoing changes in environmental regimes have enabled many theoretical innovations and given birth to complex narratives of environmental change. In significant part, existing narratives of environmental political change are framed in terms of loss and recuperation, appropriation and resistance, ignorance and enlightenment (Gadgil and Guha 1992; Lynch and Talbott 1995). Instead of local populations losing control over their resources as a result of central state policies, they can now be seen as recuperating at least part of that lost control. Rather than the state being victorious in its efforts to expropriate the valuable resources of indigenous populations, the resistance of marginalized populations can now be valorized as successful in stemming the tide of centralization. Whereas top-down policies of control and exclusion were portrayed as a product of greed and ignorance, the new decentralization of environmental policies can be attributed to a greater awareness of the need to pay attention to local variations and knowledges. Similar stories about a shift from bureaucracy to democracy, colonization to freedom, and state to community can be told. The title of a recent work, "from exclusion to participation," which refers to the involvement of local communities in environmental control, provides one way of viewing changes in environmental politics in the 1980s.[2]

But these ways of understanding and analyzing the nature of environmental politics can be enriched and supplemented. Processes around the environment always involve power/knowledges and subjectivities and are always mediated by institutions. Instead of a selective conceptual focus on "politics," "institutions," or "subjectivities" as *the* foundation on which to build an analysis of changing environmental relations, it can be more fruitful to examine how these concepts shape each other and are themselves constituted.

Indeed, each of the chapters in this book has shown the articulation among these concepts. Application of new statistical, botanical, taxonomic, and silvicultural knowledges in the late nineteenth century and the early twentieth made possible representations of forests that persist even today. But official use of new knowledges was closely tied to official recognition of the value of certain kinds of timber, the formulation of new institutional regimes, and the exclusion of social actors such as swidden cultivators. Similarly, imperatives of a better bottom line for the colonial enterprise, control over the Indian territories through reliable and fast transportation, a secure supply of raw materials needed in World War I, and bureaucratic politics within the colonial state combined in unexpected ways to lay the groundwork for the forest councils in Kumaon. The invention of new institutional forms changed the nature and possibilities of control by introducing dispersed but coordinated regimes of regulation throughout Kumaon. Indeed, even the birth of environmental subjects rested in no small measure on changing institutions and knowledges and the widespread possibilities for regulatory participation in the new institutions. These examples demonstrate that the seemingly diverse fields of social action and change denoted by knowledge, politics, institutions, and subjectivities in reality run through each other. In treating them as separable domains of human practices and scholarly analyses, we are forced to consider their articulation inadequately at best. But it is precisely in examining how these concepts and their referents make each other that it becomes possible to imagine what a new environmental politics might look like.

One may argue with some justification that in different ways scholarly contributions to environmental politics, whether before the 1970s or during and after the 1980s, are not really about environmental politics. In each period, those observing environmental politics have written, often brilliantly, either about the environment or about politics but less often about environmental politics. Paraphrasing Richard White, one might say that "it is as if trying to write the history of a marriage, one produces a biography of the wife, and placing it next to the biography of the husband calls it the history of the relationship itself!"[3]

In this final chapter of the book, I begin by looking at select writings on environmental politics that have attempted to make sense of governmental strategies aimed at environmental protection and the enactment of such strategies. My discussion examines in particular three interdisciplinary fields: common property, political ecology, and feminist environmentalism. It is by necessity in the nature of a critical review, but it helps prepare the ground for an argument about what an engaged environmental political analysis must include. My own position on the nature of environmental politics is based as much on my understanding of the unfolding of environmental processes in Kumaon and elsewhere as it is on some of the more influential recent writings about the environment. But most importantly it is articulated as an effort to craft an investigative framework that can synthesize insights from a range of environmental writings, especially as they illuminate historical-political changes in Kumaon's forests. Rather than viewing institutions, political forces, or subject-related transformations as *the* effective and sufficient loci of environmental political analysis or action, I examine how these concepts relate to and produce each other.

Environmental Political Analyses since the 1980s

The noted feminist economist Bina Agarwal suggests that "we are seeing an emerging consensus, both among scholars and among government and non-governmental agencies, that local resources should be managed by village communities" (1998: 60). Agarwal is referring to changes in state policies and scholarly writings all through the late 1980s and the 1990s that view communities and local management of resources as critical in founding the government of nature. The consensus toward which she points may never emerge, but its makings are partly visible in the explosion of studies on the local government of resources, especially those concentrating on the role of communities and communal institutions.[4] It is even more evident in the increasing number of national governments that have begun to eschew their reliance purely on coercive, top-down, centrally directed conservation policies.[5] The failure of earlier efforts at exclusionist environmental control,[6] fiscal crises of the state in the wake of the debt crises of the late 1970s and the early 1980s, the collapse of state socialism and the subsequent hegemonic status of a neoliberal orthodoxy in economic policy circles, and the availability of international aid funds for pursuing programs of decentralized governance are in no small measure responsible.

Concrete efforts to involve new actors in the government of nature have received attention from a number of scholars. I find useful the alternative routes mapped by the scholarship on the commons, writings on political ecology, and feminist environmentalists. These three approaches to environmental politics can be seen as more or less directly examining how institutions, politics, and identities affect environmental processes and outcomes. Coming into being and gathering force roughly around the same junctures in the 1980s, these cross-disciplinary approaches to environmental politics differ in important ways from their many disciplinary cousins—among them, conservation biology, environmental anthropology, environmental ethics, environmental history, environmental sociology, historical ecology, and social ecology. For one thing, compared to the environmentalist scholarship that has as its objective a new niche within the discipline, these three approaches focus on specific problems and conceptual foundations as their markers. Rather than trying to find a basis for inclusion within an originating discipline and showing how more mainstream historians, political scientists, sociologists, or anthropologists would benefit from a greater focus on environmental issues, these cross-disciplinary orientations attempt to examine the environment more directly.

Further, common property theory, political ecology, and feminist environmentalism have also been insistently problem driven and oriented toward transformations in existing forms of government of the environment.[7] Undoubtedly, environmental research that owes its primary allegiance to disciplinary formations has contributed significantly and enduringly to the insights that common property theorists, political ecologists, and feminist environmentalists have used and often built on. But, equally certainly, political ecologists, feminist environmentalists, and scholars of common property speak with a more engaged voice in discussions on forms of environmental governance and how to change exclusionary, centralized, nondemocratic control over resources. Given the overarching concerns of this book—how institutions, politics, and identities shape each other at the same time that they influence new knowledges and practices around the environment —it is fruitful to build on these interdisciplinary sites of knowledge production about the environment.

The Fields of Common Property

Writings by scholars of common property provide some of the earliest and best-known arguments in favor of self-organized government of resources.[8] Ostrom's (1990) path-breaking synthesis of the importance of common property resource regimes built on at least a decade of earlier studies (Alexander 1982; McCay and Acheson 1987; Netting 1981; National Research Coucil 1986). By beginning with a review and critique of environmentalist writings that saw the "gross concepts" (Shapiro 1989) of states and markets as the appropriate institutional avenues through which to address conservation failures, Ostrom's *Governing the Commons* (1990) articulated the stakes involved in focusing analytical attention down to microinstitutional regulation of the environment. It outlined the possibilities of community and was part of a blossoming of locally oriented studies that took as their point of departure the assumption that small groups of users could craft viable forms of environmental government.[9]

Work on the environment that uses a common property framework has burgeoned since the 1990s. Perhaps the two most important contributions of this scholarship are by now widely accepted. The first of these is the principle that variations in institutional arrangements to govern environmental goods can have a marked effect on their disposition and that successful forms include those under which users cooperate with each other to govern resources locally. Thus, there is no teleological or even deterministic logic to instituting private property and privatizing nature. A second important contribution of this literature is the recognition that concepts such as private, public, or common are too gross to account adequately for the massive variation in institutional forms that environmental subjects deploy to govern their resources (McKean 2000).

A number of scholars working on the commons have, in addition, pointed to the multiplicity of factors that affect the prospects of environmental government. It is fair to infer from the work of Wade ([1988] 1994), Ostrom (1990), Baland and Platteau (1996) and many other commons researchers that institutional variation is importantly related to environmental actions and outcomes (Agrawal 2001c). The presence of different institutional forms can have a significant influence on how environmental processes unfold.

But even a cursory examination of the literature on common property is sufficient to reveal some of the continuities in focus and, perhaps as a result of these continuities, some persistent debilities as well. For one thing, scholars of common property have tended to take institutions—qua property—as the focus of their analyses. Social practices, especially those related to environmental regulation, are for them typically the consequence of institutional transformations; institutions are seldom the visible symptoms and markers of social practices.

The focus on institutional effects is related to the origins of writings on common property. Scholars of the commons began by trying to demonstrate that common property can potentially be as efficient a solution to problems related to public goods as private ownership or state control. Though successful in this objective both theoretically and in terms of policy shifts, they have tended either to assume that distribution under communal government is more equitable or to place only a limited emphasis on questions of allocation and distributive politics. But neither strategy will be useful if the objective is to understand better the conditions under which environmental government works successfully. The preceding chapters have shown the pervasive role of political negotiations and social struggles in Kumaon in producing precisely the institutions that are the focus of study for most scholars of common property. The previous two chapters have further shown the unequal burden of seemingly equitable institutional arrangements under asymmetric social relations. Since all social relations are politically asymmetrical, it becomes crucial to understand how the effects of even seemingly equal and symmetric institutional rules fall unevenly on those subject to the rules (see Gibson 1999).

In response to criticisms posed over the past few years, scholars of common property have begun to acknowledge the critical importance and impact of a larger political-economic and social context on institutional outcomes. Contextual variables, difficult as they are to define independent of the questions being researched, clearly affect the ability of specific groups to use and govern their resources. Studies of decentralized mechanisms of environmental politics (Agrawal and Ostrom 2001), attention to the origins of commons institutions (Ostrom 1999), and analyses of heterogeneities within groups (Varughese and Ostrom 1999) are witness to this trend. But it is worth noting that even in these changing substantive concerns institutions retain their preeminent place as the objects of analysis and explanation. Thus, common property theorists may have begun to attend to questions of politics, but for them the effects of politics on resources are always tracked through institutions. How political relations or changes in the relative power of different actors affect the environment even without institutional changes is usually only inadequately considered by scholars of common property because of their primary focus on institutions.[10]

The Shifting Nature of Politics in Political Ecology

It is in the writings of political ecologists that one can find a more direct focus on questions of power and politics in relation to the environment.[11] The substantive connection and thematic divergence between political ecology and studies of common property become visible in the assertion that "political ecology is essentially a politics of the commons" (Wells and Lynch 2000: 93). Such a focus on the commons is more characteristic of political ecologists concerned with global environmental processes and the erosion of global commons. But even political ecologists who are interested in more regional or localized changes around environmental issues consider a legitimate concern of political ecology to be redefinitions of access to commons in the context of changing relations between capital and state.

Scholars in both fields—common property and political ecology—thus are concerned about the commons. Many political ecologists, for example, have written about the disappearance of the global commons and the many threats to local commons. But they differ in the nature of their concern. Scholars of common property seek to address the problem of disappearing commons by looking at rules and institutions, whereas many political ecologists focus especially on the politics inherent in the erosion of the commons and the changing forms of access to environmental resources. Especially lucid expositions of political-ecological analyses grounded in a social-historical context can be found in the works of Bryant (1996), Bunker (1985), Leach (1994), Neumann (1998), and Peluso (1992).

For many, political ecology became a recognized field of environmental studies in the 1980s. In its emphasis on politics, it is the result of the intersection of political economy and several existing fields of environmental study, especially cultural ecology, human ecology, and critical human geography (Bryant and Bailey 1997: 1–10; Peet and Watts 1996: 4).[12] In contrast to the first wave of environmental writings, which were dominated by Malthusian visions of exploding human populations and resource shortages (Ehrlich 1968; Meadows, Randers, and Behrens 1972; Ophuls 1977), early political ecologists focused on distributive aspects of resource consumption and access when examining global environmental problems.[13] When examining small-scale, more localized human-nature interactions, they often focused not only on distributive issues but also on the cultural practices that could be explained in terms of environmental concerns about stability and long-term sustainability. But from the very beginning, and in significant contrast to writings on common property, there are a wide variety of approaches among political ecologists.[14] This diversity led Peet and Watts to claim that "political ecology seems grounded less in a coherent theory than in similar areas of inquiry" (1996: 6). But their call for a coherent theory itself risks difficulties in their simultaneously expressed belief in the "natural construction of the social" and their advocacy of a poststructuralist turn in political ecology.[15]

Criticism on the grounds of methods, coherence, and systematicity notwithstanding (see Peet and Watts 1996; Moore 1996a), different approaches in political ecology shared at least three common commitments during the 1980s and the early 1990s (Bryant and Bailey 1997). The first might be seen as an insistence on questions about social marginality and access to resources. The second was the desire to investigate the political causes and effects of resource allocation. And, finally, political ecologists argued in favor of close attention to the cultural, socioeconomic, and political contexts that shape the human use and control of resources.[16] These commitments are visible even in the work of Blaikie and Brookfield, who suggest that political ecology "combines the concerns of ecology and broadly defined political economy. Together this encompasses the constantly shifting dialectic between society and land-based resources, and also within classes and groups within society itself" (1987: 17).

In the mid- to late 1990s, it was possible to identify two closely related developments in political ecological analyses: (1) a more intimate examination of politics, often through the continued use of historical and ethnographic approaches; and (2) a turn toward poststructuralist theory (Peet and Watts 1996; Escobar 1999). Claims that early political ecologists paid insufficient attention to politics or that they did not adequately theorize its role in ecological practices are less applicable to more recent work.

But my more purposive interest in political-ecological writings leads toward two specific areas of concern rather than global charges of incoherence: the first concerns the political, and the second is related to the nature of the subject that animates politics. On the one hand, the primacy accorded the political often prevents political-ecological analyses from examining how the political itself is made. The use of politics and power to explain the nature, causes, and effects of resource management and allocation casts politics as the prime mover, the cause that exists sui generis. But the exercise of power and political asymmetries—the focus of political-ecological writings—are themselves a consequence of many different processes and can be understood only historically.

In a related vein, political-ecological analyses require a more robust exploration of the politics of subject formation. It would be accurate to claim that political ecology approaches the environmental subject tangentially at best. Whether it is the land manager standing at the center of Blaikie and Brookfield's political ecology (1987: 239–40), the critic of

Western reason and the harbinger of emancipation in Peet and Watts's poststructuralist advocacy (1996: 3, 37), or the absent agent at the heart of Escobar's antiessentialist message, the question of subject formation is seldom raised adequately let alone addressed carefully. The subject is always already present in political-ecological writings.

Attempts to examine the idea of the agentive subject and theorize the emergence of ecological/environmental subjects typically take the form present, for example, in the important review of political-ecological writings in Bryant and Bailey 1997. Organizing their review through an actor-centered framework, they examine states, multilateral institutions, business interests, nongovernmental organizations, and grassroots actors. But each of these actors and their interests are represented as existing fully formed. The relationships of subjects to the environment, however, need to be examined in their emergence, not simply taken as part of a larger politics by preexisting interests. How environments, and the history of practices in relation to the environment, transform actors and interests is an enormously interesting and complex question, as chapter 6 showed. Actors work on their own interests as part of their constitution as environmental subjects. Bryant and Bailey and other political ecologists would likely concede the possibility of changes in interests over time and variations in them across different spaces. But a meaningful concession would also imply a more careful investigation of the processes whereby interests change and of the mechanisms that relate interests to social-structural locations on the one hand and to practices on the other. Also important would be attention to relationships among interests, imagination, and the production of subject positions. To pursue such a making of environmental subjects, it would be necessary to give up the concept of subjects and interests that are always already given by their social-structural locations and instead examine how they are made. Some recent writings that can broadly be included in the domain of political ecology have begun to pay significant attention to processes of subject formation (Li 2000; Moore 1998; Sivaramakrishnan 1999; Worby 2000).

The Gendered Subject of Feminist Environmentalism

The work of feminist environmentalists, like that of most ecofeminists, is founded on assumptions that connect environmentalism and feminism, often by advancing the thesis that politics and injustices around gender are closely related to and parallel those around the environment (Gaard 1993: 1; Sturgeon 1997: 23).[17] But in contrast to some ecofeminists, who have argued in favor of a spiritual (Starhawk 1982) or biological (Mies and Shiva 1993; Salleh 1984; Shiva 1988, 1994) foundation for the relationship between women and nature, feminist environmentalists are committed to investigating how economic processes, social practices, and political relations are instrumental in producing gender-related inequalities. Serious disagreements about what accounts for gendered inequalities and injustices in environmental practices and outcomes are perhaps matched only by the disagreements about how to study them. Rather than accepting a universal or essentialized relationship between women and environment, as early ecofeminists often asserted, feminist environmentalists are committed to examining gender-environment connections in a more materialist and contingent fashion (Agarwal 1994; Jackson 1993a, 1993b; Leach 1994).[18] It is not surprising that they have generated some of the more exciting and fruitful scholarship in relation to the environment.

Feminist environmentalism comprises a range of approaches, some of them claiming specific names for themselves, as is evident from a recent blossoming (Agarwal 1992; Alaimo 1994; Mellor 1992, 1997; Rocheleau 1995; Salleh 1997; Seager 1993; Sturgeon 1997; Warren 1997).[19] Apart from sharing what Warren calls the minimal conditions of an ecological feminism, there is additional common ground among these approaches.[20] Feminist environmentalists agree that gendered relationships in households, within communities, and around the environment are historically and contextually variable and socially and politically complex. In many cases, they closely examine how the burden of political-economic injustices often falls on the bodies and labor of women and the mechanisms through which such unequal burden sharing is translated into environmental degradation. Critical of the romanticized and reductive views of women in the developing world that lead to a failure to attend to political economy, feminist environmentalists seek to insert material, political-economic, and cultural processes into their analyses of gender and environment (Jackson 1993c; Jewett 2000). As Agarwal puts it, the inattention to political economy is precisely what turns ecofeminist analyses into a "critique without threat to the established order" (1992: 153). It is obvious that the at-

tempt to defend the role of excluded groups and identities on naturalized grounds consolidates a very peculiar conception of the environment (Agrawal and Sivaramakrishnan 2000). It treats the relationship
between human nature and the environment as a primordial fact, unin-
fluenced by experiences and changing social relations that may be
changing precisely because of environment-related conflicts and negotiations. It reifies contingent relationships between social identities and
environmental processes. It is inattentive to how social identities are
shaped by social practices and how individual subjects reshape themselves in response to their changing experiences of environmental government.

Two possibilities are the necessary outcome of the recognition that
there is no deterministic relationship between the interests of women
and the conservation of the environment. First, under some conditions
efforts to regulate and protect the environment can work against
women, and, second, there may be other conditions under which
women act in ways that do not further environmental conservation.
Such recognition also shows that environmental projects cannot count
on women for environmental protection as a matter of course; instead,
the extent to which women will act to conserve depends crucially on
how conservation is related to their historically constituted material
interests and the practices of which they are a part. Similarly, the assumption that women are somehow closer to nature and act as its
custodians and trustees can lead to policy designs that reserve for them
additional tasks to protect trees and vegetation without commensurate
attempts to change the political relations that marginalize them. This is
exactly what Jackson (1993c) shows when she suggests that attempts at
conservation often relegate women to marginal positions of power and
simultaneously increase their labor requirements. The absence of
women from decision-making positions in most forest councils in Kumaon is a fact that supports this argument. Recall the unequal burden
environmental regulation and enforcement placed on women and
lower caste members in Bhagartola, as discussed in chapter 5.

Although feminist environmentalists have successfully contested the
easy conflation of the category of woman with environment and forcefully pointed to the regressive potential inherent in such naturalized
relationships,[21] they have been less successful in examining the role of
power in producing women as environmental subjects. Rather, the ex-

ercise of power is what excludes already constituted women from pos-sibilities of participation and access to, or control over, environmental resources (Agarwal 1994). Feminist environmentalist analyses can be greatly strengthened by a closer treatment of how differentiated en-vironmental experiences create gendered subjects or how they affect environmental outcomes. This would also, however, require giving greater primacy to practice rather than the social identity category of gender. Indeed, the privileging of gender as the primary basis for inves-tigating environment-related injustices or inequalities has another im-portant consequence. It means that, although strategies for under-standing gender can be extended to other social identities, the work of feminist environmentalists focuses on subject positions other than those related to gender inadequately.

Elements for an Environmental Politics

It is evident that the three sets of writings described above have contrib-uted importantly to better analyses of environmental actions and out-comes. They have helped frame the terms of environmental discourses and at the same time guided the thinking and training of a generation of young environmentalists. They have been especially effective because they each have clearly articulated foci of analytical interest—institutions for scholars of common property, politics for political ecologists, and the gendered subject for feminist environmentalists.

The specific focus of each of these cross-disciplinary interventions in environmental politics can also be enriched, however. Attention to in-stitutions by scholars of common property often means that they un-derplay variations in subjectivities or consider politics and knowledges only in their relationship to institutions. But surely there is an entire domain of political practices and changes in subjectivities that affects resource use and government and cannot be approached by a primary focus on institutions. Indeed, the emergence of new knowledges, often in intimate relationships with institutions, has the potential to affect the bounds of what can be imagined as the environment and actions in relation to the environment.

Political ecologists consider politics far more carefully than do schol-ars of common property. Yet their work, like that of feminist environ-mentalists, often consolidates a particularly narrow conception of the environment. They view the environment as an arena in which conflicts such as those between elite and poor, state and community, or outsider and local unfold (Bryant 1996: 221–25; Hughes 2001). Indeed, these oppositional terms for portraying politics and conflicts are often structurally interchangeable. Elite are part of the state, or at least have intimate connections with state actors. Communities, and the poor who are their members, are locally situated (Colchester 1994; Klooster 2000; Lynch and Talbott 1995). In consequence, even acute analyses of political conflicts and environmental histories are often constrained to reach particular conclusions. Such constraints on the analytical imagination are especially evident in assessments of recent shifts in environmental politics in which communities and decentralization have emerged as important watchwords (Klooster 2002). Thus, some scholars valorize communities and decentralization (Ghai 1993; Gurung 1992). Others see government-local partnerships as Trojan horses that facilitate the maintenance of the status quo (Gauld 2000; Hill 1996; Munro 1998). In both cases, the argument often turns into the assignment of credit and blame. Depending on the initial assumptions, communities, states, or markets can conveniently be chosen. Of course, some acute work has recently begun to question the easy conflation of communities with resistance or states with power (Braun and Castree 1998, 2001; Darier 1999; Li 2001; Moore 1996b; Peluso and Vandergeest 2001; Sivaramakrishnan 2000).

Although political-ecological writings attend insistently to politics and in some cases to institutions, they have only recently begun to explore questions of subjectivity and knowledge carefully. How people understand the environment and relate to it, how new knowledges about the environment shape such understandings, and how changing institutions, politics, and subjectivities play a role in ecological practice need greater elaboration and analysis. Especially important for further investigation is a better sense of how understandings of the environment change over time, producing new environmental subjects.

It is certainly true that in contrast to political-ecological research and common property theory, a range of scholars writing about indigenous peoples and gender pays greater attention to questions of identity (Diamond and Orenstein 1990; Merchant 1980, 1990; Shiva 1988). These scholars have focused especially on the argument that "women" and "indigenous" agents should be viewed as guardians rather than ignorant bystanders in relation to the environment. This focus on the subject contrasts interestingly with the almost missing subject in political ecology and its nearly complete absence in writings on common property. However, feminist environmentalists focus on the making of subjects mainly in relation to gender. By drawing on perceptive accounts of subjectivity and agency (Butler 1989, 1993; Haraway 1989, 1991), it may be possible for feminist environmentalists to move away from analysis in which gender remains a proxy for discrimination.

Foucault and Environmental Politics

An approach to environmental politics that builds on the contributions of common property theorists, political ecologists, and feminist environmentalists needs to be especially attentive to the production of new power/knowledges, institutions, and subjectivities, not just to their role in affecting environmental outcomes. In developing such an approach, I draw selectively on Foucault's later work, especially where he introduces and discusses the idea of governmentality. But the application of Foucauldian insights requires more than selectivity. It also necessitates supplementation and reconfiguration. My purpose in this reconsideration is less to locate precisely what Foucault did or failed to do and even less what he could or should have done (Stoler 1995). Instead, my objective is to use Foucault's work as a source of possible provocations leading toward a framework for environmental political analyses.

Foucault's initial use of the term *governmentality*, or *governmental rationality*, was aimed partly to address those critics of his work who saw it as focused too directly on the micropractices of power—as, for example, in *Discipline and Punish*—and attending too little to macropolitical relations.[22] By now a large number of scholars have adopted the term governmentality and applied it to problems of government, especially in liberal democracies, in an effort to show how politics has changed in these contexts over the last two centuries (Foucault [1978] 1991).[23] This scholarship has focused especially on modern manifestations of power in arenas such as education, social economy, insurance, risk management, welfare, criminality and police, space and architecture, and security, among others.[24] Its goal is to understand and describe how modern forms of power and regulation achieve their full effects not by forcing people toward state-mandated goals but by turning them into accomplices. The very individuality that is supposed to be constrained by the exercise of power may actually be its effect. In this sense, these analyses of modern government trace the profound transformation in mechanisms of power over the last two centuries (Foucault 1978). The power of death over subjects implicit in sovereignty now exists together with its counterpart: power that can exert "a positive influence over life, that endeavors to administer, optimize, and multiply it, subjecting it to precise controls and comprehensive regulations" (137).

Governmentality outside Western Modernity

The explosion of the problem of government, according to Foucault ([1978] 1991), occurs when one of the functions of state power becomes the administration of life.[25] The regulation of life as a goal of government raises with particular intensity the problems of "how to govern oneself, how to be governed, how to govern others, by whom the people will accept being governed, how to become the best possible governor" (87). However, these questions cannot be answered fully without the invention of specific techniques that allow the treatment of both the individual and the social body as complete, self-contained entities. In one case, the science of the individual can be applied to shape his or her actions; in the other, the science of statistics can be applied to the life of the population.[26]

Although Foucault notes that the emergence of population shifts the emphasis of statecraft from sovereignty and discipline to effective government, issues related to sovereignty and discipline do not disappear ([1978] 1991: 102): "Sovereignty is far from being eliminated by the emergence of a new art of government" (93). Rather, a new set of problems comes to be a part of the legitimate domain for the exercise of state power. In this manner, the need to govern effectively complicates the exercise of sovereignty (Kuehls 1993: 141). It is also important to note, with Hacking's important historical study, that statistics is not just about discovering the rhythms of populations and demography. "Statistics has helped determine the form of laws about society and the character of social facts. It has engendered concepts and classifications within the human sciences. . . . It may think of itself as providing only information, but it is itself part of the technology of power in a modern state" ([1981] 1991: 181).

The two forms of power in social relations that simultaneously govern and take care of the entire social body as they also try to reach out and care for each individual are precisely what Foucault has in mind when he calls the characteristic property of modern government a government of "all and each."[27] The government of all and each necessitates paying meticulous attention to economy in political practice.[28] In Kumaon, when such economy could not be effected by centralized government, state officials and local elite together devised new strategies of decentralized government that helped achieve economies in both political and social regulation. The real strength of the state lies in population—in the strength and productivity of all those who are a part of the state: "Police is a science of endless lists and classifications; there is a police of religion, of customs, of health, of foods, of highways, of public order, of sciences, commerce, manufactures, servants, poverty. . . . [It] seems to aspire to constitute the sensorium of a Leviathan. . . . Police government works by the means of specific, detailed regulation and decree."[29] Certainly this view holds for the government of the environment. From the detailed classifications of Indian landscapes to the different types of forests that we encountered at the beginning of British rule and the tabulation of different forms of criminality that necessitated regulatory innovations in Kumaon, the production of lists, tables, numbers, and rules is a critical part of modern government.

The increasingly detailed elaboration of the means by which governance takes place is accompanied by an expansion and increase in the number of social arenas that potentially require government. It is interesting to note that every new form of governmental action in relation to the social simultaneously underlines the distinction between state and society and also seeks to bind together more closely the working of the state and the social. Government is possible because the state and the society can be conceptualized as separate entities. At the same time, regulatory actions covering ever broader aspects of life demonstrate the difficulties of an autonomous existence of the social and reemphasize the social as the raison d'être of government. It is such regulatory action, or strategies of government, that produces the effect of both the community and the state. The point, then, is not to dispense with the use of *state* or *community* but to recognize and emphasize the politics involved in their contingent production.

These dual, "mutually–in–tension" aspects of governmental actions become possible because of the way they target individuals. It is changes in the practices of individual persons, with each a member of society and all of them collectively constituting the social, that are the object of regulation. Solutions to problems associated with some aspect of the social—high birth rates, low levels of industrialization, deforestation, and underdevelopment—require changes in individual behavior. Governmental strategies achieve their effects, to the extent they do so, by becoming anchors for processes that reshape the individuals who are a part and the object of governmental regulation.[30] By attending to practices, it becomes possible to see how institutions, politics, and subjectivities together comprise different technologies of government.

The need for modern government arises out of several processes that highlight population as an entity with its own regularities, cycles, and effects. As Foucault writes: "The welfare of the population, the improvement of its condition, the increase of its wealth, longevity, health, etc." become the object of government ([1978] 1991: 100). The knowledge of population becomes available through the field of political economy, and its regulation takes place through different apparatuses and institutions of security in such a fashion that it leads to thorough governmentalization of both what we call society and what we call the state. Refinements in the sciences of demography and statistics reveal those macrolevel features of a population that one must know in order to govern in a rational and conscious manner.[31] The construction of steadily more elaborate lists and tables about the qualities of the population, the effort to know the rhythms and regularities of the social, the launching of the processes that "make up people" (Hacking 1993), and the governance of these people are thus all part of governmentality.[32]

It can be argued that the remarkable body of work that has used governmentality to trace the history of changing forms of governmental authority and practice in the West has led to a focus on governmentality as a primarily liberal phenomenon.[33] However, it is governmentality as an analytical optic that is obviously relevant to other places and historical periods. Treating governmentality as an analytical construct exposes its potential usefulness in investigating the nature of institutionalized power outside of Western modernity.

In general, technologies of government may be characterized as being founded on some combination of knowledges, regulations based upon these knowledges, and practices that regulations seek to govern. But institutionalization of new strategies of power and regulation is also accompanied by changes in conceptions of the self, a point that often eludes scholars of institutions and regulations alike. Not only is such an interpretation of governmentality a useful means of investigating those forms of power that seek to shape conduct, but it also bears the promise of connecting disparate domains of analysis: those concerned with the political-economic aspects of institutional and organizational shifts and those focusing on transformations in subjectivities.

Even in Foucault there are suggestive indications for an analytical rather than only a historically specific treatment of governmentality (Dean 1999). The most important lead into such an argument stems from Foucault's views on power and its relationship to the subject.[34] His apparently neutral accounts of technologies of power have led many to complain that he is normatively confused and has no basis for a critique of the social phenomena he describes (Fraser 1989; Philp 1983; Rorty 1984; Taylor 1984: 152–53).[35] But such complaints, as Patton 1989 shows, miss the force of Foucault's arguments. They misinterpret how he uses the concepts of power and freedom. Foucault's views on power constitute many innovations, but what is critical to a discussion of subject formation is his thesis that power is not just something that operates negatively on preconstituted subjects. Rather, one of the prime effects of power is how "certain bodies, certain gestures, certain discourses, certain desires come to be identified and constituted as individuals" (Foucault 1977: 98).[36]

The relationship between subject formation and power rests for Foucault, then, on an utter refusal to view power simply as the ability of a person to make another do something,[37] and is predicated on a positive conceptualization of both freedom and power.[38] Power is not just about the ability to constrain certain kinds of actions, peoples, or outcomes; it is as much about the possibility of producing them. The distinction between positive and negative freedom is widely accepted (Taylor 1979; Rose 1999), even when the bearer of positive freedom is conceptualized differently by people such as Taylor and Berlin.

Positive freedom is the capacity to act in certain ways. The creation of subjects depends on their exercise of power in the service of a goal.

But subjects do not form themselves by exercising power abstractly, as if in a vacuum outside of history. It is in adopting certain actions, gestures, and desires over time that their practices produce the effects of their subjecthood. It is not surprising, then, that in addressing the constitution of subjectivity Foucault talks about practices and knowledges that have a historical dimension: "New techniques for examining, training, or controlling individuals, along with the new forms of knowledge to which they give rise, bring into existence new kinds of people" (Patton 1989: 264).

It is possible to distinguish three modes of subject formation, in each of which power in a positive sense plays a critical role. Subject creation occurs through scientific inquiries that focus on and help identify particular types of subjects as their targets—for example, the productive and laboring subject of economics, the speaking subject of philology and linguistics, or the subject/citizen dichotomy of normative political theory. Subject formation also takes place through disciplinary practices that instantiate distinctions such as those between the sane and the insane, the sick and the healthy, or the author and the reader. Finally, the ways a human being turns himself or herself into a subject by following certain practices and modes of thought constitutes a third, practical mode of making subjects.

These three modes of subject formation are interdependent. The first two are examples of what Hacking calls nominalism (1986). They intersect with the third in mutually reinforcing ways. For Foucault, the realm comprising different practices of sexuality is a prime example of the third mode of subject formation (1982: 208). The imagining by some people that they are members of a given national community (and the refusal by others to imagine the same sense of belonging) or the self-fashioning of some as environmental subjects (and the continuing lack of concern about the environment in others) exemplify other possibilities. In the argument I have advanced, practices of specific subjects are the location where relationships between institutions and power/knowledge and imagination and subjectivity come together. They are a basic mechanism on which subject formation rests. The adoption of particular practices at any point in time is itself a prior result of variable combinations of politics, institutions, and existing subject locations, of technologies of government.

It should be obvious that in my argument different modes of subjectification are not necessarily specific to a particular place or historical period.[39] A similar argument about other elements comprising governmentality can also be advanced: knowledges aimed at regulation, and the targeting of social practices through regulation, emerge as part of an analytical framework for understanding the conduct of conduct in the domain of environment. Governmentality is a useful lens through which to examine political regulation and programs of governance. Different forms of governmentality are a consequence of transformations in two arenas: in the nature of the elements that comprise governmentality—power/knowledge, institutions, and subjectivities; and the kind of activities and contexts that are the focus of attention: development, welfare, education, economy, health, or advanced liberal democracies versus colonial rule. This book has focused on a particular domain of regulation—the environment in the colonial and postcolonial periods.

The analytical conceptualization of governmentality can be further underlined by taking a closer look at one of its other distinctive aspects. The exercise of power under liberal governmentality is distinguished in Foucault by the importance accorded to knowledge. Government is the right disposition of things, so as to produce a convenient effect. The right disposition of things can be known only through careful investigation and the deployment of techniques and procedures that yield knowledge about the needs and desires of a population. Chief among these techniques and procedures are those guided by reason.

However, a raft of literature now points to the difficulties involved in separating different forms of regulation and knowledge production into two categories—one guided by modern reason and another characterized by irrationality.[40] It is equally difficult to cleave to a view that forms of knowledge and regulation based on reason came to fruition only under liberal government, displacing earlier knowledges, which were dependent on unreason. It is more defensible to suggest that regulations are always dependent on some form of knowledge rather than mere caprice. It is the nature of various forms of knowledge that is always potentially under dispute.[41]

Nor are regulations claiming to serve the interests of those being regulated unique to modernity. Plato's Laws argue that a legislator or lawgiver should know the nature of his people and the conditions under which they live before he makes laws for them: "The art of governing people is rational on the condition that it observes the nature of what is governed" (Foucault [1981] 1988: 149). In the third century B.C., Kautilya gave the following instructions to Chandragupta Maurya in India: "In the happiness of his subjects lies the king's happiness, in their welfare his welfare. He shall not consider as good only that which pleases him but treat as beneficial to him whatever pleases his subjects" (Rangarajan 1987: x). And Foucault cites Saint Thomas, who explained that "the king's government must imitate God's government of nature" ([1981] 1988: 149).

The Government of Environment

A raft of new work that uses some derivation of governmentality to analyze political and policy-related innovations, especially in non-Western locations, has begun to sail into view (Mitchell 2000; Prakash 1999; Scott 1995, 1999; Stoler 1995). This scholarship on governmentality is a testament to the fecundity of an approach that interrogates some of the most cherished positions in the heated (and dated) debates on the relationship between state and society.[42] A careful consideration of the concept of governmentality provides useful tools to sidestep the state-society distinction and the debates swirling around it. Indeed, recourse to governmentality has the potential to demonstrate how even those works that question the idea of the state by talking about state formation still remain wedded to a common conceptual architecture that takes states and societies as its basic building blocks. Instead of examining the boundaries and definitions of the state and society, an analysis of governmentality orients attention toward the concrete strategies to shape conduct that are adopted by a wide range of social actors and how these different actors collaborate or are in conflict in the pursuit of particular goals. As a result, it becomes possible to move around familiar questions about the nature of the state and the extent to which states shape social processes. On the other hand, even the best-developed analyses of state formation and state-society relations, because they take the distinction between states and societies as self-evident rather than historically and politically produced, remain subject to this limitation.

Writings on governmentality can illuminate and track the uncertainties and unexpectedness of new forms of government in at least four ways. First, they prompt analyses of how problems that require government (the conduct of conduct) come into being rather than accepting unquestioningly the existence of problems. For example, analyses of development and environmental conservation that use the optic of governmentality do not just identify causes of underdevelopment and environmental degradation; nor do they just propose solutions. They also ask when and for what reasons these processes came to be identified as problems that merit a particular style of analysis and resolution (Herbert-Cheshire 2000). In late-nineteenth-century Kumaon, the problem of environmental degradation was identical for forest officials with the problem of indiscriminate logging by private contractors, uncontrolled use of timber and vegetation by shifting cultivators, and extraction of fodder and grazing by village residents. The solution required the exclusion of most human influences except logging as long as logging occurred under the auspices of the forest department.

Second, instead of taking power as the fixed property of some agent(s), it becomes possible to examine concretely how power is generated by and located in different strategies of government. No particular agent or person can then be seen as being located in a permanently more powerful political position vis-à-vis another. Instead, one can begin to ask why some strategies of government work in certain ways and with what effects. In Kumaon, for example, the revenue department was able to prevail against the forest department by finding unexpected aid in the protest strategies chosen by local residents. The power of specific practices of involvement in regulation and enforcement is the basis for the emergence of new subject positions. The power of statistics made possible the adoption of similar governmental solutions to the problem of significant biophysical diversity in the landscapes that the forest department sought to domesticate.

Third, recourse to governmentality as an optic also orients attention toward the careful study of the techniques, forms, and representations of knowledge that are related to new means of governance. Statistics, maps, numerical tables, and their collation in specific formats can become the basis for producing new forms of knowledge that make some actions seem naturally more appropriate than others as an invaluable aid to the process of government. Similarly, the monitoring of council-managed forests produces specific and intimate knowledge about village residents that then allows the forest council to select those forms of threats and sanctions that are most likely to prove effective.

Finally, studies of governmentality bring to the forefront questions about the relationship between government and self-construction. If the literatures on institutional analyses, public policy, and the state treat the process of subject formation and identity change as lying outside their legitimate domain, an examination of governmentalization is about integrating institutional and other social changes with changes in subjectivities. The foregrounding of questions about these relationships forces analysis to search for resources that would allow at least the beginnings of answers.

Contrast the possibilities of an approach that takes governmentality seriously with the position Ferguson ([1990] 1994) adopts in using Foucault to deconstruct development processes in Lesotho. Although he explicitly deploys the term *governmentality*, he uses it to signify what he sees as the proliferation of oppressive state power and institutions.[43] Early in his analysis, he describes governmentality as the principle according to which "the main features of economy and society must be within the control of a neutral, unitary, and effective national government, and thus responsive to planners' blueprints" (72).[44] In this version of governmentality, isolated and identified as bureaucratic proliferation, the dichotomous classification state/society continues to form the analytical foundation. The term is equated with expansion of state control over social processes. Ferguson's valuable point about how arguments portraying the objective need for state expansion transform development into an apolitical, technical endeavor is bought at the cost of a Weberian gloss that invokes centralization as the phantom haunting state-led development efforts. Ferguson's reading of governmentality through the lens of bureaucratization forgoes the opportunity to explore the multiple forms of conduct in a polity and ignores questions about how subjects of development come into existence and with what consequences. It is important to explore the different technologies of government that states pursue and their relative prospects of success. And, although it is fair to suggest that development discourses colonize subjects, surely they do not colonize all subjects. But the portrayal of the developmental state as an expanding bureaucratic entity that comes to control its population misses the chance to examine when, through what means, and with what prospects of success those subject to policies use them as part of their struggles to improve their life chances (cf. Pigg 1992). Equally lost to view are the changes in subject positions that take place together with changes in policy.[45]

Environment-related changes in Kumaon and a closer reading of Foucault suggest a somewhat different interpretation of governmentality in relation to environment. *Environmentality*, the term I find useful in this context, refers to the knowledges, politics, institutions, and subjectivities that come to be linked together with the emergence of the environment as a domain that requires regulation and protection. Regulation always demands new knowledge. But the production of new knowledges is intimately connected to the shaping of practices and human subjectivities in relation to the environment. Since politics always implies interactions and negotiations, it also always signifies the mutual constitution of fields of action related to regulation and practice. These considerations suggest that, although it may be linguistically and analytically convenient to parse *environmentality* as a combination of four elements, the working out of environmental politics implies concurrent changes in them. It is difficult to imagine the emergence of new forms of knowledge without changes in political relations and institutional arrangements and the development of new ways of thinking about the object of knowledge—the human subject.

Environmentality refers, then, to a specific optic for analyzing environmental politics instead of denoting a particular form of it. Specific forms of environmentality depend on the characteristics of the elements constituting it. Shifts in the nature of knowledge, politics, institutions, and identities lead to new forms of environmentality by definition.

The substantive chapters in this book have examined how each of the elements comprising environmentality have changed over the last century and a half in Kumaon. Take, for example, the two chapters in part 1. They explored at some length the use of numbers and statistics to organize the vast amount of new information that had become available about India's vegetation and landscapes since the end of the eighteenth century. In the first instance, the strategic and commercial needs of the East India Company formed the impetus for the production of this new information. Surgeon civil servants of the company were important fulcrums for the initial taxonomic advances in botany. But from the mid–nineteenth century on the increasing importance of timber revenues, the sheer quantity of new information, and the belief that more systematic exploitation of the subcontinent's vegetation was in order led to institutional innovations in the form of new departments of forestry in all the major provinces. These institutional changes went hand in hand with the training of a new cadre of forestry officials, who saw themselves as the guardians of India's vegetation and timber. In a sleight of mind that challenges the imagination, they portrayed themselves as guardians of forests at the same time that levels of timber exploitation reached unprecedented heights. Part and parcel of this mental legerdemain was the portrayal of other actors who might be interested in forests—timber contractors, merchants, shifting cultivators, peasants, and revenue department officials—as ill intentioned or ill informed (or both) about preservation of the environment.

Even this brief recapitulation of the discussion related to the production of forests in India (and Kumaon) should be sufficient to show that the generation of specific kinds of knowledges is contingent on and goes together with important political, institutional, and subjectivity shifts. An understanding of how one of the elements comprising environmentality changes requires a consideration of its relationships with other elements. The chapters in part 2 of the book illustrate this point equally well. Each chapter examines the triad of politics, institutions, and subjectivities in the context of altered knowledges about forests. But to facilitate exposition within each chapter the focus remained on changes in a specific element and on the role other facets of environmentality played in relation to the changes under consideration. Thus, chapter 4, on "dispersal of regulation," provided an in-depth investigation of how intracommunity institutions changed in Kumaon between the 1920s and the 1990s. But the making of new institutions at the level of the village cannot be understood without attending to the ecological practices that underwrote them, negotiations over their character and precise makeup, and the distributive conflicts they generated.

Institutional regulation, ecological practices, and subject formation are related to and depend on various forms of knowledge. But a central and particular feature of the knowledges that became a part of environmental regulation from the 1860s onward was their genesis in and invocation of expert authority. Even more crucially, new ways to produce knowledge, through statistics and numbers, were combined with the claims to expert authority on which regulations depend (and which shape practices). Problems of the social body, as perceived by state officials, were instruments in the production of knowledges that then became the basis for new policies designed to address deficiencies in existing strategies of regulation. The yoking of multiple forms of regulation to claims of expertise and truth was equally crucial to technologies of government that sought to create and rationalize regulations around forests in Kumaon from the 1860s onward. Indeed, it is possible to suggest that joining them has been critically important for environmental regulation elsewhere as well and continues to perform a significant normalizing task in programs of environmental government. The combination of knowledge and regulation is critical, finally, in the processes of self-formation that can be viewed as the consequence of the interactions between regulation and situated practice. The implementation of new regulations in Kumaon and the creation of new practices went hand in hand with changes in the human subjects who were the target of new regulations, even when the changes that transpired were not what government had intended.

Although this discussion of environmentality follows closely the historical and political discussion around forests in Kumaon, it should be obvious that the concept can also be applied to the politics of other environmental concerns. Specific governmental strategies are often a result of attempts to regulate and can have as their inspiration a whole range of phenomena, from observations of existing practices to assessments of inherent dynamics of natural resource "systems." Consider as an example the phenomenon of global climate change. Increasing fears about the likely upward shift in average global temperatures have prompted widespread discussions about the creation and allocation of carbon quotas as a form of regulation to limit the production of carbon dioxide. The reduction of emissions and different strategies of carbon sequestration, many hope, will positively influence outcomes related to global climate change.

In discussions about the allocation of carbon quotas, the differences among the positions taken by various states are reflected in the way they cluster around various proposals to limit the production of greenhouse gases. But each proposal is an attempt at regulation of existing practices (even though no international actor has the power to enforce international agreements). The practices that regulations seek to transform are themselves a result of multiple causes, and their alteration depends on the abilities of humans to transform nature in accordance with their desires. In the regulation of carbon emissions, some countries see themselves as guardians of the global environment, especially those in northern Europe and other parts of the developed world. Others portray themselves as needing to pursue higher levels of development and demand compensation in the form of transfers of funds and new technology. They argue that such transfers of wealth and technology are necessary in part because they have been the victims of centuries of high levels of production of greenhouse gases in the North. It is in such tensions between regulations and practices that the ground of politics and political knowledge is generated.

Conclusion

Environmentality, as the term has been used in this book, constitutes a way to think about environmental politics. It attends carefully to (1) the formation of new expert knowledges; (2) the nature of power, which is at the root of efforts to regulate social practice; (3) the type of institutions and regulatory practices that exist in a mutually productive relationship with social and ecological practices and can be seen as the historical expressions of contingent political relationships; and (4) the behaviors that regulations seek to change, which go hand in hand with the processes of self-formation and struggles between expert- or authority-based regulation and situated practices.

In opting for environmentality as the optic to examine the long process of changes in environmental politics, institutions, and subjectivities in Kumaon, this book has insisted on the importance of considering these concepts and their referents jointly. It is through the attempt to examine them together that it becomes possible to see how new technologies of environmental government emerge, the problems that they are devised to address, and the degree of success they enjoy. Technologies of government, as Zygmunt Bauman might say, "cast human reality as a perpetually unfinished project, in need of critical scrutiny, constant revision and improvement" (2000: 229). They are about the effort to change existing "modalities of being"—for nature as well as humans. As they were applied in Kumaon in the twentieth century, they were reflected in governmentalized localities, regulatory communities, and environmental subjects.

A focus on technologies of government and their application helps undermine the tendency to view institutions, power, or subject locations as the unquestionable starting point from which to gain an understanding of environmental change and politics. Instead it encourages attention to the processes through which these concepts are consolidated and naturalized. It directs analysis to the interdependent constitution of these three seemingly foundational concepts in environmental studies and thereby makes the familiar contingent.

Nature's government today may have come a long way from what Saint Thomas had in mind. But for all that it is equally closely tied to the government of humans—visible in the changing forms of regulation and subject formation in Kumaon and equally identifiable in global discussions about changing climate and declining biodiversity. We gain a richer awareness of environmental politics by beginning to trace the connections among power/knowledges, institutions and subjectivities. It is toward such an end that this book has proceeded.

Notes

Chapter 1 Introduction

1   The leaders I met included Sundar Lal Bahuguna and Chandi Prasad Bhatt, the two prominent personalities associated with Chipko. For a recent careful study of the movement, its leaders, and their different strategies, see Rangan 2000. See Mawdsley 1998 for a thoughtful reflection on how Chipko has assumed an iconic status in conservationist arguments globally.

2   The most careful and detailed review of official forestry records from this period is likely Shrivastava 1996.

3   Official reports and surviving accounts of villagers' actions suggest that many fires were deliberate protests against state interventions. *Incendiarism* is the term officials used to denote fires set on purpose. Note how it fails to signify villagers' own interpretations of why they were setting fires.

4   The amount of forest land that was fired rose to 272,000 acres. Of the 819 offenses that were detected, 395 were classified as incendiary (Guha [1989] 2000: 52, 107, 116).

5   The increasing importance of timber in the hills, especially as the commercial and strategic value of pine for railway sleepers and turpentine was realized, led the forest department to adopt increasingly draconian measures to appropriate forests ever more comprehensively (ibid.).

6   The nature and history of these regulations are subjects of discussion in chapters 2 and 3. See also Baumann 1995; Chaturvedi and Sahai 1988; Dangwal 1996, 1997; Farooqui 1997; and Shrivastava 1996.

7   Villagers' protests were similar in nature to many peasant movements that occurred throughout the subcontinent over the period of British rule (Hardiman 1992).There was some sense of being wronged, some organization, and an identifiable target. The difference was that the threats to existence in many of these peasant revolts was partly attributable to natural disasters compounded by the callousness of the government, whereas in Kumaon the causes of protests lay almost entirely in the efforts of the state to separate villagers from their means of livelihood.

8   I use the terms *custom* and *customary* loosely, but with some sense of their complex history and the ambiguities involved in their use in the wake of

arguments about "inventions of tradition" (see the essays in Hobsbawm and Ranger 1983). In my use, the terms simply suggest that the new laws paid little attention to how forests were connected to rural livelihoods. Their passage, motivated mainly by the desire to raise revenues for the state, created a legal framework that sought to restrict and restructure prevailing livelihood practices.

9 For a discussion of similar colonial interventions, see Peluso 1992; and Sivaramakrishnan 1999. The rise of illegality in Kumaon was made possible in part by internal changes in the forest department. Between 1911 and 1921, the number of employees in the department rose from thirty-five to nearly one hundred (KFGC 1921). The growth in the number of detected violations and the ensuing convictions between 1911 and 1933 can thus be seen as a result of both new regulations and higher levels of enforcement. It is also true that reported violations were no more than a very small proportion of what the colonial state would have liked to prevent. But the social, political, and economic costs of attempting to detect all violations would have been prohibitive.

10 For a more detailed discussion of the committee's recommendation, see KFGC 1922. See also Agrawal 2001a.

11 The figure is based on data compiled from the annual reports of the forest department in United Provinces. Guha ([1989] 2000, tables 2 and 4) reports some of these data but unfortunately incorrectly, as is evident from a comparison with the actual figures in the annual reports.

12 If the modern state is to be seen as a "principle of *organization*" (Lloyd and Thomas 1998: 3), the means of implementing this principle was what registered a shift beginning in the 1920s in Kumaon.

13 In the specific design of the rules, state officials also drew on existing practices of local forest regulation in some Kumaoni villages (Shrivastava 1996). The reliance of colonial conservation on existing local practices is also noted by Grove (1995) and helps question the presumed watershed that colonial rule is often taken to mark. On the general question of whether colonial rule in South Asia can be viewed as constituting a complete break with the past, see the revisionist arguments in Bayly 1988; Stein 1985; and Washbrook 1988. Chatterjee, in contrast, sees colonial power as a distinctive "rule of colonial difference" in which the colonized and their practices are represented as inferior (1993: 19).

14 For Rose (1999: 52), technologies of government shape conduct and are dependent on the assemblage of "lines of connection among a diversity of types of knowledge, forces, capacities, skills, dispositions, and types of judgment." He goes on to include "forms of practical knowledge, modes of perception, practices of calculation, vocabularies, types of authority,

forms of judgment, architectural forms, human capacities, non-human objects and devices, inscription techniques and so forth." His idea of assemblage is a useful metaphor, but for my own analysis his comprehensive listing of the constituent features of such assemblages is both too comprehensive and too specific. It is inadequately attentive to the relations among these aspects or to how they these assemblages might be used more broadly to study specific instances in which technologies of government are formulated and applied.

15 The term *environmentality* has been used by Luke (1995, 1997), who views it as an attempt by transnational environmental organizations to control and dominate environmental policy and activities around the world but especially in developing countries. See also the collection of essays in Darier 1999. My use of the term is indebted to Luke for the coinage but is different in both intent and meaning. I attempt to examine more insistently the shifts in subjectivities that accompany new forms of regulation rather than seeing regulation as an attempt mainly to control or dominate.

16 Sivaramakrishnan's *Modern Forests* (2000) contributes significantly to environmental studies that take the idea of Foucauldian governmentality seriously. Indeed, he was one of the first scholars of environment in South Asia to examine the possibilities of the concept of governmentality, especially in relation to strategies of power that contribute to the idea and reality of the state.

17 Their influence, rippling out over the past seven decades, is now being undermined by the introduction of the joint forest management rules in Uttarakhand (Sarin 2002).

18 The manner of villagers' control over themselves is visible in the records many of the forest councils maintain. These records show that the protection villagers ensure, the policing tasks they undertake, and the monitoring mechanisms they have created have led to an imperfect but ultimately far more comprehensive mechanism of regulation than that exercised by the forest department.

19 Also present were a few lower-level officers from the district revenue administration and the forest department and representatives from some local nongovernmental organizations (NGOs).

20 These rules were created by forest council officials in conformity with the guidelines laid down in the Forest Council Rules of 1931.

21 I also met and talked with residents in eight villages that did not have forest councils (see chapter 6).

22 A perceptual split between ordinary village residents, who depended on forests for the conduct of their daily lives, and forest council members,

who sought to regulate forest use, was present in the responses of most villagers. Although the members of the councils relied on forest products themselves, they were also seen as agents trying to translate the will of the state as it was expressed more than fifty years earlier in the Forest Council Rules of 1931.

23 See Sivaramakrishnan 2000. Although his use of *modern forests* is primarily a reference to colonial Bengal, forms of modernity are always in flux. The term is evocative of the emerging regimes of regulation in Kumaon as well.

24 I use *transnational* in the sense of Mann (1988: viii), who contrasts *international* (characterizing relations between national economies and classes) and *transnational* (relations proceeding across the boundaries of states).

25 The obvious popularity of idioms of resistance in the wake of Scott's landmark study (1990) and in critical response to his analysis of ideas of negotiation and engagement would be an all too easy trap into which one might place the history of forested environments in Kumaon. For some studies that elaborate on these idioms, see Colburn 1989; and Haynes and Prakash 1992. The two-volume study by Comaroff and Comaroff (1991, 1997) also discusses metaphors of resistance, but it examines changes in subjectivities under the Marxist-inspired term *consciousness*. Mitchell (1990) constructs a careful critique of Scott's arguments about weapons of the weak and resistance.

26 But see Rose ([1989] 1999), who discusses psychological procedures and their impact on how subject positions change. He provides a careful treatment of how human conduct in general is shaped through mechanisms of power. See also Butler (1997), who points out the absence in Foucault's writings of any mechanisms through which power produces subjects.

27 Two groups of people that were affected most directly by British efforts to assume control of vast territories that contained trees were timber merchants/contractors and indigenous tribal people who practiced some form of shifting cultivation. The latter were also the groups for which British foresters reserved their greatest opprobrium (Baden-Powell 1893; Ribbentrop 1899; Stebbing 1922–26).

28 Contemporary discussions in the nineteenth century and the early twentieth defended the need for forestry regulations and state control over forests. They consistently remarked on the shortsightedness of policies that judged private working of forests to be adequate for conservation and regeneration. Stebbing's influential account of early forest history repeatedly underscores how private interests can never be harnessed to public objectives where forests are concerned because of the cupidity of timber merchants and contractors (1922–26: 75, 141, 249). See also Ribbentrop

1899: 74–75. Nor were such discussions particularly Indian. For similar arguments about West African forests, see Unwin 1920 and Stebbing 1937; for European forests, see Fernow 1907; and for forests in China and East Asia, see Shaw 1914.

29 The clearing of forests and expansion of agriculture were long seen as synonymous with the spread of civilization. Harrison cites Vico, who wrote: "This was the order of human institutions: first the forests, after that the huts, then the villages, next the cities, and finally the academies" (Vico 1968 quoted in Harrison 1992: 246). Harrison goes on to suggest that "forests mark the provincial edge of western civilization" (1992: 247). In Kumaon itself, for the first fifty years of British occupation forests were seen as inexhaustible and their clearing was considered desirable and necessary for agricultural pursuits (see chapter 3).

30 Ribbentrop, the third inspector general of forests, cites Brandis's initiative approvingly, noting that he "correctly foresaw that if the people of the country could ever be brought to plant Teak in their shifting cultivation, this would be likely to become the most efficient mode of artificially reproducing the tree . . . the prophecy has become true" (1900: 73).

31 The sheer scale and spread of government efforts to make a regulatory partner out of communities makes enumeration repetitive at best, but it is worthwhile to point to a few cases (drawn from FAO 1999). In Bolivia, after the enactment of legislation on decentralization and peoples' participation, provincial plans were developed and 144 local institutions were involved in protectionist agreements. In Brazil, 412 projects have been launched to support traditional communities. Among African countries, in Guinea a service to coordinate NGO interventions was established in 1994 and forest user groups were recognized in 1997. In Senegal, people's participation and the involvement of local communities in the governance of forests is widespread. In Sierra Leone, village forestry associations have been encouraged, and in Sudan the government has made agreements with many local communities and councils. In Asia, Nepal has developed mechanisms for leasehold and community forestry, and the state envisages transfers of almost all forests in the middle Himalaya to community organizations. In Indonesia, a Village Forests Development Program is under way. In Papua New Guinea, joint implementation agreements with local governments and communities have been established. In Thailand, community forestry is being widely implemented. In Vanuatu, most of the forest lands belong to community groups. These are only samples of the large variety of agreements and initiatives under way in different countries.

32 Inden (1990: 131–42) provides a careful analysis of several important

historical writings on Indian communities, among them those by Marx
and Maine. His examination is intended mainly to elucidate the construction of the "Indian Village."

33 As paid, direct appointees of the state, village headmen and other revenue officials had little leeway in interpreting the responsibilities they
were supposed to discharge.

34 The vast literature on common pool resources and local governance
provides many guidelines, and not always the same ones, for shaping the
use of resources. See Ostrom 1990 for a rigorous introduction to writings
on the commons and Crook and Manor 1998 for a comparative analysis
of programs of decentralized development.

35 Writings by members of the Subaltern Studies Collective provide fine
lessons in reading official archival evidence against the grain. Guha and
Spivak 1988 constitutes a useful introduction.

36 Mamdani 2001 offers a strong argument relating political institutional
changes to those in the identities of the colonial and the postcolonial
subjects.

37 Although indirect rule as a practice of government first developed in
India (Fisher 1991), it is better known through its practice in sub-Saharan
Africa (Lugard 1926).

38 See also Ribot (1999), whose study of indirect strategies of environmental management in Africa builds on Mamdani's arguments. The specific
logic of rule, decentralized or centralized, does not have to depend only
on despotic force. It is arguable whether current efforts at decentralization are best interpreted as creating subordinated populations that resemble subjects in Mamdani's sense. The nature of rule depends on the
mix of ingredients that come together to shape practices. Among these
ingredients would be mechanisms for the selection of rulers, the relationship between rulers and laws, and the nature of recourse to procedures of mediation and adjudication.

39 It is worth pointing out that decentralization of specific responsibilities
and powers is a feature of business organizations as well. Miller tells the
story of the Volvo automobile plant in Kalmar, Sweden, where teams of
up to twenty workers complete major subassemblies without monitoring, control, or incentives from higher management. These second-order
tasks are delegated to the team as well. Hazards of team production are
familiar to most economists and organization theorists, and Miller explains the high morale and similar assembly times in Kalmar, as in other
plants, by referring to theories of repeated games and property rights.
But the full explanation depends on "*mutually reinforcing psychological
expectations*" (Miller 1992: 180, my emphasis).

40 For a discussion of these themes in the works of Marx, Durkheim, and Weber, see Rose 1999.

41 For examples, see the studies in Elster 1985.

42 For reviews and research on decentralization in different arenas, see Crook and Manor 1998; Fox and Brown 1998; and Grindle 2000. Putnam's (1993) modern classic on decentralized institutions in Italy also speaks to the importance of decentralization in development.

43 Since changes in subjectivities may be crucial in other arenas of decentralization and government reforms as well (Dean 1994b; Pavlich 1996), it might be fair to suggest that the forest councils of Kumaon illustrate the new distribution of power that decentralized regulatory communities inaugurate. They may be historically distinct from a larger contemporary political shift, but they are analytically tied.

44 This shift in the fortunes of marginal groups and local communities is what Ribot (1995) is referring to when he uses the phrase "from exclusion to participation." Others have similarly written of the transformation that environmental politics has undergone over the last three decades (Western and Wright 1994).

45 For comparative figures on revenue yields of forests under different provincial departments, see Stebbing 1922–26. For similar figures on revenues in World War I, see Champion and Osmaston 1962.

46 In many ways, decentralization processes connected with the Joint Forest Management Program (JFM) are today pushing the Indian forest department in directions that Kumaon's forestry officials traveled in the 1920s. Village communities, through JFM committees, have gained some managerial powers in nearly a fifth of India's forests (Khare et al. 2000). However, many observers of JFM conclude that it is quite modest in its decentralizing thrust (Kumar 2001; Sundar 2000).

47 See, for example, Eckholm 1976; and Wilson 1992.

48 For some representative views that defend state intervention in environmental conservation and management, see Hardin 1978; Heilbroner 1974; and Ophuls 1973. Others portray privatization and markets as the preferable choice (de Alessi 1980). In particular, resource economists have advocated private property solutions to the problem of environmental degradation. For critiques of state policies to conserve resources, see Ascher 1999; and Repetto and Gillis 1988. See Baland and Platteau 1996 for a formal demonstration that there is no difference in the efficiency characteristics of private property and managed common property or community management of natural resources.

49 In joining governmentality and the environment, I depart substantially from existing treatments, which view environmentality as primarily a

means of supervision by international networks of environmental organizations (Luke 1995). Where governmentality is concerned, scholars of development and environment often use it to signify new forms of domination and expansion of government (Ferguson [1990] 1994). Even Akhil Gupta's careful *Postcolonial Developments* (1998), which draws heavily on Foucault to discuss governmentality and transnational organizations, interprets the concept primarily as a technique of systematization, surveillance, and imposition of new orders instead of also attending to the positive aspects of power involved in government and the production of new subjects.

Part I Power/Knowledge

1 Whereas the French had intensified interventions on land and in forests within France in response to their need for naval timber, the British government displaced its resource shortages to the colonies (Albion 1926; Bamford 1956; Grove 1995: 57–60). British military successes on the Indian subcontinent in the late eighteenth century and the early nineteenth, especially after the defeat of Tipu Sultan in 1792, made the displacement of the demand for timber to the colonies much easier.

2 Malabar forests had served as an important source of timber for shipbuilding by Indian merchants, local rulers, and the British for more than thirty years (Grove 1995: 390–92). For example, in 1799 ten thousand mature teak trees were floated down from these forests. Demand had increased to such an extent that it had even become necessary to import teak from Rangoon (Stebbing 1922–26).

3 Similar effects of the imperial machine's voracious appetite for timber were felt in New England when the lumber trade spurred colonization in the mid–eighteenth century and when trade in ports such as Portsmouth and Falmouth depended almost entirely on a single item: timber (Albion 1926: 274–75). I am thankful to James Scott for leading me to this source.

4 According to Albion's remarkable study of the role of timber in British naval power: "The most successful attempt to relieve the [British] timber shortage by foreign shipbuilding came from the use of teak in India. . . . England armed dozens of ships built in India for her expeditions against Ceylon, Java, Manila, and the Moluccas, and into the Red Sea between 1795 and 1800 . . . [and] had it not been for the use of these improvised warships, her control of Indian waters would have been doubtful. . . . The choicest timber lay in the southwestern part of India, in the region known as the Malabar Coast" (1926: 365–66).

5 Watson's charge extended to all forests in Malabar and Travancore (Madras) that were not claimed as private property. His main task was to ensure adequate supplies of teak for the government.

6 Although merchants were allowed to fell timber with the permission of the conservator, they could not market it. In effect, this meant that they had to follow prices fixed by the conservator if they wanted to get any return at all from their investment in harvesting timber (Ribbentrop 1899; Stebbing 1922–26).

7 Brandis, the first inspector general of forests in India, said about the conservatorship: "The first attempt at forest management was a great mistake, an act of injustice which cannot be condemned too severely" ([1897] 1994: 97). No small role in the abolition of the conservatorship was played by Sir Thomas Munro, the governor of Madras. Munro died in 1826 after a zealous career in which he assiduously applied the principle of eliminating intermediary authority between the company and its subjects, but not before taking strong exception to the activities of Watson (Stein 1989: 59–60). Richard Grove (1995: 397) attributes to Munro an idealized and highly inaccurate view of indigenous landholding and suggests that he was unaware of the influx of timber traders and land-grabbers in the region, who had little concern for conservation. To boot, Munro was a firm believer in laissez-faire policies and argued that a free market in timber, instead of government regulation, was necessary (Ribbentrop 1899: 65).

8 As Miller observes in his study of colonial conservation in Brazil, timber was a resource unmatched in the early nineteenth century by any other natural product: "Timber was not only the steel, aluminum, plastic, and fiberglass of past ages, but the oil, coal, and gas as well." Speaking of the uses to which it was put, he remarks that "structurally, timber had no competitor, as it is the only naturally occurring substance that exhibits tensile strength. . . . By weight, most woods are stronger than steel. . . . The wooden ship was the period's highest expression of material culture and the great tool of European expansion, domination, and commerce without which world history would be another story altogether" (2000: 4).

9 Although John Nef saw the sixteenth and seventeenth centuries as the "age of timber" (1966) Lindqvist argues for the importance of timber even until later: "Not only did mining consume large amounts of wood; so, too, did potash plants, tanneries, glassworks, saltpeter works, train-oil works, lime production, and other industries that rely on the forests for fuel and raw materials. Domestic demands included fuel for heating houses and drying grain and malt, and timber for houses, fences, ships, carts, barrels, and agricultural implements" (1990: 301).

10 Many scholars have tried to answer the question of how colonial interventions changed the use of forests by calling colonial forestry policies
"scientific forestry," "production forestry," or "rational forestry," as if
what preceded colonial rule was unscientific, unconcerned with production, or irrational. Others have tried to argue that colonial practices did
not introduce anything qualitatively or fundamentally different in the
management and exploitation of Indian forests. Two of the figures associated with these polarized positions are Ramachandra Guha (1983, 1989)
and Richard Grove (1995). See also the introduction to Grove, Damodaran, and Sangwan 1998 and the epilogue to the second edition of Guha's
The Unquiet Woods (2000). See Rangarajan 1994 for a review of the significant points of debate between Guha's and Grove's earlier writings.
Richard Tucker touches on several of the more important themes in this
debate (1983, 1988a, 1988b).

11 Khare et al. (2000) review the basic outlines of the Joint Forest Management Program and its most recent achievements. A vast literature already exists on this program. For a brief introduction to this literature
and a comparison between JFM and some other decentralization policies, see Agrawal and Ostrom 2001.

12 For important exceptions, see Sivaramakrishnan 1996 and Pratap 2000.
For two earlier studies from outside South Asia that place contemporary
forest histories in a longer time frame, see Peluso 1992 and Bryant 1996.

Chapter 2 Forests of Statistics

1 In his remarkable history of measurements, Kula (1986: 13) contrasts our
pervasive reliance on counting and numbers with an aversion to it in the
not so distant past. "To count and to measure [was] sinful," he wrote.
"Among the Czechs, at the end of the eighteenth century, a belief was
prevalent that a child under six years of age would cease growing, become stunted, a 'measureling,' if the cloth intended for his shirt or outer
garment was measured. . . . In Macedonia, at the end of the nineteenth
century, peasants would not eat what had been measured for fear of
developing a goitre. In the Vladimir *gubernya*, at the beginning of the
second half of the nineteenth century, peasants were inimical to the
practice of calculating what they had harvested." Such doubts about
measurements were also to be found in nineteenth-century India. See
Prakash (1999: 42–43).

2 I am indebted to Hacking (1999: 6) for this distinction in his analysis of
social constructivist studies in the 1980s and 1990s. I use *forests* to refer

both to a physical entity and to the idea of that entity, preferring this strategy over that of always using a different sign for each concept. Concerns about readability have prompted my choice, and I trust that the different meanings of the term are clear in my usage from the context.

3 Useful thematic introductions to writings on Indian environmental history and politics are available in Grove, Damodaran, and Sangwan 1998 and Agrawal and Sivaramakrishnan 2000.

4 I am especially indebted to the important work of Arjun Appadurai (1993) and Talal Asad (1994), on numbers and power, and the thoughtful research of Theodore Porter (1986, 1996) and Ian Hacking (1999) on the history of statistics and the social construction of ideas and objects. Scott (1998) comes closest to the arguments in this chapter in discussing strategies of legibility and simplification that are a part of authoritarian state projects in high modernity. For a useful introduction to the history of statistics, see Rose 1999: 197–232. Recent studies in the field have insisted on the significant role of numerical representations in constituting and constructing what they seek to represent (Alonso and Starr 1987; Hacking 1986, 1990; Porter 1996). Other studies have fruitfully examined "statistical social construction," to coin a somewhat awkward phrase, in specific domains (Appadurai 1993; Hopwood and Miller 1994; Patriarca 1996).

5 For Himachal Pradesh, see Saberwal 1999; and Singh 1998. For Uttar Pradesh, see Dangwal 1996, 1997; and Srivastava 1996. For the Central Provinces, see Anderson and Huber 1988, which undertakes the study of a contemporary project; Prasad 1994; Rangarajan 1996; and Sundar 1997. For Bengal, see Sivaramakrishnan 1999. For western India, see Skaria 1999. For Madras, see Philip 1996. And for Burma see Bryant 1996.

6 The literature on each of these themes is immense. For some recent studies on gender and environment that do not take these as fixed categories, see Agarwal 1994; and Gururani 1996. For social movements, see Rangan 2000; and Weber 1988. And for a discussion of the reciprocal constructions of the relationship between the agrarian world and the environment see Agrawal and Sivaramakrishnan 2000; and Dangwal 1998. This new scholarship has shown the regionally variable relationships between colonial officials and those subject to colonial rule; the development of official forestry practices, existing use and management patterns, and local distribution and growth patterns of vegetation; and the diverse articulation between the intentions of officials as expressed in written documents and plans and the lived experience realized in the processes of creating forests.

7   Sangwan (1998) questions much earlier research on colonial science, which saw it as an inferior collaborator in the larger enterprise of European scientific research (Basalla 1967; Kumar 1980; McLeod 1975). Other studies have demonstrated the importance of European models of the environment and forests to colonial policies and the diffusion of scientific ideas within the colonies (Rajan 1998a; but see also Sivaramakrishnan 1999).

8   Stebbing (1922–26) has written perhaps the most comprehensive imperial history of forests in British India. Earlier efforts by Brandis ([1897] 1994) and Ribbentrop (1900) were formative in their influence but far less complete. Champion and Osmaston (1962) have made a useful attempt to extend a state history of forests that follows the same plan as that of Stebbing's earlier volumes.

9   This taxonomic exercise is accompanied by strict generalizations based on the relationships between the categories that the science of forestry constructed and connected to categories from other sources. These generalizations brook no exceptions and must have been invaluable to budding foresters, giving them a bedrock of certainties about the forests they would be managing. Consider the following, for example: "As a rule, tropical evergreen forests which grow on metamorphic rocks are richest in species, while those occurring on the sandstones and other sedimentary less-altered rocks are poorest in this respect" (Stebbing 1922–26: 42–43).

10 Stebbing was prodigious in his output. Quite apart from volumes in the more autobiographical vein, he wrote three books on forestry in Britain (1916, 1919, 1928) and another on colonial forestry in West Africa (1937).

11 This is just the beginning of Stebbing's book. After invoking the all-India terrain of forests, he goes on to describe in massive detail what happened once the British began to concern themselves with the wealth that forests contained. He organized his account by period and within each period by major political divisions. Numerous contributions from scholars, ethnographers, administrators, travelers, statisticians, and tabulators form the basis of his history. Ranging over more than 1,900 pages, he describes personalities and their actions, state policies and their origins, and new technologies of exploitation and their consequences. It is an invaluable source with which to begin tracing the narratives that constitute the colonial knowledge of Indian forests, even though Stebbing affects a staunchly official stance throughout and seldom provides bibliographic references.

12 Interestingly, after Brandis's retirement in 1881 the two foresters who followed him as inspector generals—W. Schlich (1883–88) and B. Ribbentrop (1888–1900)—were also trained in Germany (Ribbentrop 1900).

13   The combination of the objectives of revenue maximization and sustained yield required assessments of the availability of the most valuable timber in the forests; protection, preservation, extension, and consolidation of existing forests; and rationalization in harvesting and marketing trees. According to Lowood, three basic principles linked the desideratum of a regulated, systematically managed forest to the methodological focus on measurement and calculation: "minimum diversity," "the balance sheet," and "sustained yield" (1990: 333). He describes how these principles were translated into the practice of forestry in Germany once the objective of foresters became systematic management to enhance supplies.

14   Brandis's method could be used for any tree species, and it continued to be used with some modifications for more than half a century, especially in forests stocked with a high density of timber-bearing trees and reserve forests. In view of the importance of survey work, the government of India created a separate Forest Survey Branch in 1872, which functioned independently until 1900, when it was absorbed into the Survey of India as a branch of that organization (Ribbentrop 1900: 132–33; Eardly-Wilmot 1906: 3). All forest surveys were placed under the superintendent of forest surveys, and the general direction of the branch was transferred to the surveyor general of India in 1904.

15   The method Brandis used is reminiscent of the one developed by Johann Gottlieb Beckmann of Saxony. Beckman employed a team of assistants equipped with nails of various colors to represent different size classes. As the group walked through the forest, members estimated the size class of trees they passed and marked the individual trees with nails of appropriate colors. At the end of the operation, the unused nails were counted and subtracted from the original supply, yielding a count of trees in each class. That number could be converted into a volume of wood using appropriate multipliers. I am indebted to Lowood (1990: 325–26) for this description.

16   The description below is based on heroic assumptions that equate area with value. Over the course of the nineteenth century, foresters in Europe developed a number of other strategies to smooth fluctuations in revenues by basing the principle of extraction on factors such as the volume or value of timber (Lesch 1990; Lowood 1990).

17   Despite the obvious difficulties involved in realizing such a scheme in practice, Heske informs us that "as early as 1359, the city forest of Erfurt in the German territories was divided into as many annual cutting areas as there were years in the rotation, and one such area was cut each year. . . . A simple regulation of yield on the basis of area was also carried out in

the Mannsfield forests in the Harz in 1588 and at about the same time in the Miltenberg forest" (1938: 22).

18 For a comparative history of forestry in different European states, see Fernow 1907. Fernow's work is itself based on other published materials for specific countries. Kirkwood (1893) provides a collection of papers dealing with forestry in the nineteenth century in different European countries. For information on French forestry, see Woolsey 1920. Woolsey (1917) also discusses the effects of French influence in colonial forests in Tunisia and Algeria. Heske (1938) describes the beginnings of forestry management in Germany, dating it to the Carolingian dynasty (A.D. 751–911). He traces planting, clear-cutting, and the regulation of timber harvests through selection felling to the mid–fourteenth century.

19 Brown (1883) provides the complete translated text of Colbert's ordinance and also places it in the context of earlier forest ordinances in France. Some of these proclamations date to the thirteenth century. See also Sahlins 1994: 52–55. Whited's later history of forestry in the nineteenth century in the French Alps (2000) confirms the patchy effectiveness of Colbert's ordinance.

20 Comparing the estimated 1,200 indigenous tree species of India (in an area half the size of Europe) to the 158 species in Europe, Brandis wrote: "The forester finds himself bewildered by this overwhelming variety of forest vegetation. Few can attempt to acquire a knowledge of all these species, but a large number of the more important kinds the Indian Forester must know, if he is to do his work" (1897: 130). Brandis's estimate of 1,200 indigenous species of trees was later revised upward to more than 5,000 woody species by Troup. Of these, he estimated 2,500 to be tree species. He described and reported on nearly 550 of these species by their uses and technical characteristics (1909).

21 Schlich, in the preface to the first volume of his *Manual of Forestry* argued that British foresters "must set to work and collect statistics based on home experience" (1906: vi). See also the second and third volumes (1910, 1911). The manual was published between 1889 and 1895 (Rajan 1998a: 347–49).

22 Later, in 1881, Brandis provided a different estimate: about two million teak trees over 6 feet in girth scattered over 2,400 square miles of prime forest (1881b: 5).

23 Writing about how to improve forest administration in British Burma, Brandis noted that "the chief object of forest conservancy in Burma is to ensure the permanent production annually of a sufficient quantity of teak and other valuable kinds of timber. . . . Teak may justly be called the prince of woods, . . . Teak in the London market maintains a price which

is higher than the price of any other timber, and which is surpassed only by mahogany and one or two other furniture and fancy woods" (ibid.: 1). For Brandis, although mahogany or other woods might surpass teak in price, it was only because of their cosmetic and fancy value.

24 Figures 2 through 4 are based on statistics reported in the government of India's *Annual Return of Statistics relating to Forest Administration* (see, e.g., GOI 1930, 1941), which was published as a part of the *Review of Forest Administration in British India*. Earlier volumes were published under the authorship of the inspector general of Indian forests, but after the first twenty-five years of publication the forest department stopped providing a report each year. Instead, it published only statistical information annually. The textual reports were published every five years, yet another indication of the importance attached to numbers over text.

25 Brandis (1897) provides a fascinating glimpse into the political calculations that led to the establishment of a separate forest department in India. The alternative was to let the revenue department also manage forest lands. See also Guha 1990.

26 I am indebted for the expression "forests of statistics" to Hacking's "avalanche of numbers" ([1981] 1991: 187).

27 The three fundamental features of Linnaeus's classification were abstraction, numeration, and artificiality (Lesch 1990: 74–82). Each classifiable entity was fixed formally, beyond the accidents of time, history, or spatial particularities. The clarity and utility of Linnaeus's classification system, based as it was on the essential sexual reproductive features of plants, came at the price of empirical intuition (Lesch 1990; Foucault 1970).

28 Other notable publications from around this time include William Roxburgh's (head of the Calcutta Botanical Gardens) *Plants of the Coromandal Coast* and Robert Wight's *Illustrations of Indian Botany*.

29 For a review of such early descriptions in Bengal, see Sivaramakrishnan 1999. Grove's (1993) account of the role of surgeon-naturalists in imparting an environmentalist rather than purely economic concern to early conservationist advocacy in southern and western India also contains references to several travelers' accounts. See also Grove 1998b. Stebbing's (1922–26) discussions of reports on forests between 1810 and 1850 by such people as Clementson, Connolly, and Gibson in Malabar; Wallich in Burma and the northwest; Blair as the district collector in Canara; and Jervis (a member of the Military Board of the East India Company) in South India more generally also bear out the claim that few of these reports contained much quantitative information on forests or their condition.

30 The range and number of monographs and other publications are too

vast to be listed in any representative fashion. See Brandis 1897 (130–32) for a selected list of titles.

31   This number ignores the land in "native states," which was governed under very similar forestry practices. In 1904, an additional 12,500 square miles were being managed as reserved or protected forests in Kashmir, Mysore, Travancore, Jodhpur, Baroda, and other princely states (Eardly-Wilmot 1906: 30–34).

32   In a number of years, for example, the annual reports disclose earlier misclassifications of land, inaccuracies in the amount of land considered to be under the control of the forest department, and corrections of past figures on breaches of forest law or valuations of forest produce (Eardly-Wilmot 1907: 2).

33   In 1908, the 140,000 acres (225 square miles) of plantation land were just about 0.1 percent of the nearly 235,000 square miles of land classified as forest.

34   The *Quinquennial Review of Forest Administration* for the years 1924–29 reports: "It was hoped in Madras, by means of modern American methods, to extract and utilise very large quantities of valuable timbers, but the final result of this work was to prove that this extensive exploitation was not justified either by the stand of timber in the forests or by the possibilities of satisfying markets" (Rodger 1930: 6).

35   Prior to the Nilumbur teak plantations, the British had begun planting trees along the Western and Eastern Jumna Canals in 1820–21 and 1830–31, respectively. The chief species planted along the Western Jumna Canal were sissoo, toon, kikur, sal, and teak, and the revenue derived from the sale of wood from these plantations more than covered the expense (Stebbing 1922–26: 201–2). Other efforts were far less successful. Sivaramakrishnan (1999) explains how several attempts to create plantations led to no perceptible improvements in Bengal.

36   These rules, "a wonderfully good set of prescriptions" for the time, were so successful that in 1860 there were nearly 400,000 surviving plants and the plantation forest witnessed regular planting and thinning operations (Balfour 1862).

37   The following discussion of working plans is based on a selection of plans created to work five forest areas: the South Kabirwala and Mailsi Reserved Forests in Punjab (1899); the Kangan Range in Hazara district, Punjab (1901); the Murree-Kahuta Forests of the Rawalpindi Division, Punjab (1901); the Topla and Kuilikapah Forests in Balaghat, Central Provinces (1901); and the Uanaungmyin, Kaing, and Palwe Reserves in Pyinmana Division, Upper Burma (1902).

38   Existing vegetation in an area being considered for working plan pre-

scriptions was often referred to as crop, and the plans provided information about the state of the crop, the distribution of various species in it, and injuries to which it was vulnerable.

39 Brandis is explicit in delineating what he considered progress in forestry. On the subject of publishing more sophisticated descriptions of Indian trees, he asserts: "When forestry has made more progress in India, when successful systems of regenerating forests of the northwestern and the eastern Himalaya have been established, when the effects of fire protection upon Teak and other trees have been determined by series of comparative valuations surveys in different districts, when yield tables showing the amount of timber production per acre per annum of the principal kinds under different circumstances have been prepared, and when the chief enemies, insects and fungi, of these species are more fully known, then it will be time to publish complete and practically useful books in each province" (1911: viii).

40 Pickering (1995) examines the old adage "science proposes, nature disposes" in the context of high-energy physics. He focuses specifically on the sources of resistance in experiments that test the validity of theoretical predictions and the ways in which scientists make accommodations to such resistance by modifying their theories, models, conjectures about the apparatus that is used in an experiment, and the working of the apparatus itself.

41 Grover (1997) studies the production of timber and its role in the colonial economy in the Himalayan Punjab. She also studies how the meaning of timber changed under the British.

42 See Hacking 1999 for illuminating studies of the reciprocal relationship between classifications and the nature of the objects or entities being classified.

43 The process I am describing is analogous, in an even more defined fashion, to Cohn's description of the production of "cultural identities" in the nineteenth century. Cohn argues that to see the process of cultural change as a kind of by-product of a historical experience whose major thrust has been political and economic is to miss some of the significance of what has happened. According to him, when Indian intellectuals tried to think about their culture and identity they "in some sense made it into a 'thing': they can stand back and look at themselves, their ideas, their symbols and culture and see it as an entity. What had previously been embedded in a whole matrix of custom, ritual, religious symbol, a textually transmitted tradition, has now become something different" (1987: 229).

44 For a discussion of the classical view of a model and its use to represent

natural phenomena, see Lesch 1990: 82–84. Foucault's (1970: 138–45) description of a system, with the Linnaean taxonomy as the prime example, is valuable for the insights it affords about the epistemological arbitrariness but practical usefulness of the processes of selection that ultimately yield any system and its constituent units and relationships. A model can be taken as a product within a system that is based on the use of numbers or signs to represent the features of a phenomenon or entity.

Chapter 3 Struggles over Kumaon's Forests

1 Rangan (2000) discusses the different narratives that have been woven around the Chipko movement and how it has become a part of environmentalist folklore.

2 For two interesting, somewhat different perspectives on the relationship between nature and human interventions, see Dove 1992; and Merchant 1980.

3 Although there are few accounts of how humans governed forests prior to the nineteenth century, Champion (1923) constructs an absorbing general narrative about how human activities likely shaped the distribution of major tree species in the Kumaon Himalaya in the period for which we do not have written records.

4 The difficulties of measuring land and demarcating cultivated and uncultivated land constitute one piece of evidence about the lack of government over forests in the period before British rule. Throughout the Chand period, rural land was measured in terms of the amount of seed necessary to cultivate a given tract. According to Atkinson: "No estimate even of the area of the waste and forest land was ever made by the former governments" ([1882] 1973: vol. 3, 467). Since this varied depending on the fertility of the land, the Gorkhas tried to introduce a more uniform system of measurement. The introduction of uniformity proved to be so costly that they abandoned their efforts.

5 Three types of taxes were common: *kath-bans* (on wood and bamboo), *kath-mahal* (on *catechu*, a type of dye), and *ghikar* (on cattle). Ibid.: vol. 1, 845–46.

6 By some estimates, nearly 200,000 hill men were sold into slavery or bondage in the markets on the plains to meet the high tax demands of the Gorkhas (ibid.: vol. 3). However, Traill fixed his revenue assessments at the same level as had the Gorkhas.

7 Under the land revenue settlement, which came to be known as Traill's San Assi settlement, all land was measured, boundaries of cultivated and

forested lands were demarcated, and the domain of control for private holdings was designated as the cultivated land. The settlement made no provisions for the conservation of forests. Despite some concerns about the declining availability of teak, the impression of the inexhaustibility of forests was not peculiar to Kumaon administrators. Stebbing, reviewing the progress of Indian forestry policy in 1921, wrote the following about the early years of British rule: "The great continent appeared to hold inexhaustible tracts covered with dense jungles. . . . The early administrators appear to have been convinced that . . . in many localities forests were an obstruction to agriculture. . . . The whole policy was to extend agriculture and the watchword of the time was to destroy the forests with this end in view. . . . This was a transitory period, but enormous destruction to valuable forests was the outcome. . . . The spread of railways at a later period brought the matter to a head. . . . The Keynote of our interest in the Indian Forests between the years 1796 and 1860 may be said to have been their exploitation for timber" (1922–26: 82–83).

8   Bailey went on to say that it was only wholesale clearance and the difficulties encountered in procuring railway sleepers that made the state realize the necessity of forest protection (1924: 189–90). Responding to Bailey, whose observations were about the forests of the United Provinces in general, another forest officer added: "The extensive clearance of forests . . . referred to by Mr. Bailey are but a small portion of the whole to which the present day destruction of the Kumaon and Garhwal forests may be added. The rate at which these hill forests have been destroyed was fittingly indicated in the recent debate on supplies in the Local Council, by a reference to an old man who now had to walk 40 miles for his plough wood, whereas in his youth he could obtain it in his village" (Benskin 1924: 492).

9   According to the *Gazetteer of Nainital*, the contractors had "uncontrolled liberty to cut where they pleased, with the result that large number of trees were felled and for want of transport were left lying in the forests. To such an extent was this reckless felling carried out during this period that for several years after the control of the forests was taken in hand by the Commissioner, the energy of the officials was directed towards extracting the timber thus left by the contractors" (Neville 1904: 19).

10 Trevor and Smythies (1923: 3) paint a similar picture: "The forests were treated to an orgy of destruction in many districts. . . . After the mutiny of 1857, the great expansion of the railways developed an enormous demand for timber for sleepers, and all the accessible forests of the province were depleted of their best timber trees for the production of

sleepers and other uses." Chaturvedi, another member of the Indian Forest Service, remarked: "The first half of the nineteenth century witnessed an exploitation of forests on a scale hitherto unprecedented in the history of these provinces . . . an almost callous destruction of forests took place . . . [and] extensive tracts of forests were sacrificed for the cultivation of tea in Kumaun and Garhwal" (1925: 357).

11   According to Walton (1911: 11) Traill's proclamation restricting the felling of sal led to the "first forest reserves in Kumaon" since they were thereby excluded from leases granted by the state.

12   In both Bombay and Madras, which witnessed some of the earliest efforts to govern forests, revenue department officials also managed forests for much of the first half of the nineteenth century (Fisher 1885: 586).

13   For an in-depth discussion of the rivalries between the forest and revenue departments in Kumaon between 1890 and 1925, see Shrivastava 1996. His argument draws its inspiration in part from Guha's (1990) discussion of the differences between centralizers and decentralizers in the making of the Indian Forest Act of 1878 and the Madras Forest Act of 1883. See also Farooqui's discussion of the disagreements among British officials about the necessity to classify large areas of land as reserved forests (1997: 49–57).

14   More systematic management in Kumaon can be dated to the 1870s. Smythies argues that in 1876 "forest conservancy was systematically started in some of the hill forests of Kumaon" (1911: 54). Although many of the practices that led to systematic state control were initiated in the 1860s, far-reaching implementation of these and some of their refinements took an additional decade.

15   The gigantic elephant creeper could attain such dimensions that a single plant, "with its spreading limbs, like some vegetable octopus, [could] sometimes cover the tops of the trees over a quarter of an acre of densely grown forest. The suppression and extermination of these natural enemies is therefore one of the most important points of forest conservancy and one of the chief factors in the future well-being of a timber-producing area" (Walton 1911: 15).

16   Pant wrote that "between the property-grabbing zeal of the revenue officers and the exhortations of experts of the forest department, the rights of the people were ground down to bring forth the above notification. By this notification, all the forests and waste lands of the districts of Almora, Naini Tal, and Garhwal not forming part of the measured areas of villages or the reserved forests were declared to be protected forests under Section 28 of the Indian Forests Act" (1922: 39).

17 Forest department officials also tried not to be unduly harsh in the punitive measures they imposed. In 1915–16, the average level of fines on rule breakers in Kumaon and Garhwal was less than 20 percent of the level of fines in other parts of the United Provinces. The exact figures are approximately Rs. 1.4 for Kumaon and Garhwal as opposed to more than Rs. 7 for the rest of the province (GOUP 1916: 6).

18 Part of the reason for this drastic increase in the area that the forest department brought under its control must be sought in the shortages created by World War I. The shortage of shipping because of the war brought into "high relief the dependence of India, for *coniferous* timber, on foreign countries" (Smythies 1917: 166). In India, the Kumaon region was potentially one of the most important areas for the production of softwoods.

19 The most popular journal of forestry in India at the time, *Indian Forester*, was filled with articles and research reports that suggested links between greater rainfall and the existence of forests, and the role of forests in preventing floods, recharging subsurface water, and increased soil moisture. For representative articles, see some of the papers in a single volume of the journal: 37(3–4): 119–30; 37(7): 354–64; and 37(9): 477–88.

20 See Rangan 1997 for a discussion of how differences over the definition of access to forest products by local populations were increasingly a source of tension between the forest and the revenue departments.

21 Baden-Powell, for example, argued: "First as regards the people. They are ignorant as we have seen of the practical truths established by forest science, the more so as they are blinded by a short-sighted idea of their immediate interest. *All forest conservancy is therefore necessarily disliked.* . . . [A] real cause of unpopularity of forest conservancy . . . arises from the fact that the people continue to adhere to their own notion as to the proprietary right in the forest, while the Government declaration on the subject is, and has been for years past, at variance with such notion" (1876: 3, 4).

22 Reporting on Indian agriculture at the end of the nineteenth century, J. A. Voelcker noted that "the forest department by its intervention has stopped in a good measure the work of destruction and has . . . ensured a continuous revenue to government" (1897: 135).

23 Forest officials often complained about the limited funds available for improvement activities in forests. Indeed, compared to forest management in Prussia and France, which respectively spent 1.8 and 0.9 dollars per acre on the upkeep of their forests, the Indian forest service spent no more than a pittance, 0.076 dollars per acre (Anonymous 1912a: 224–38; 1912b).

24 A number of other writings attest to the profitability of the forest department. See, for example, Anonymous 1912b; Blaschek 1912; and Fisher 1885.

25 The coolie system refers to three types of exactions: coolie *begar*, which referred to forced labor without remuneration; coolie *utar*, which carried an obligation of minimum wage payment; and coolie *burdayash*, which referred to the appropriation of articles such as food, fuel, and fodder for officials, soldiers, hunters, surveyors, tourists, or their animals (Pathak 1991: 261).

26 Shekhar Pathak has undertaken the most thoroughgoing study of the various types of forced labor arrangements in Kumaon, their institutionalization under the British colonial state, and people's protest movements against them. He presents a summarized English version of his study in Pathak 1991.

27 Hewitt had made similar remarks in 1908 at a durbar in Bareilly (cited in Rawat 1991: 288).

28 This classification system followed the recommendations of a conference of concerned district and forest officials in 1911 (Shrivastava 1996: 204–5).

29 A 1982 survey in Almora district indicates that 57 percent of the fodder needs of cattle are met from common forest lands (Jackson 1985: 137–38).

30 Quoting Colonel Ramsay, the commissioner of Kumaon, Baden-Powell said that the Kumaon villagers "'owned their jungle in a way before we came'; and so when we recognized their proprietary right in the cultivated land, the people acquired 'a certain right in the use of the forest'" ([1892] 1974: 310).

31 It may be useful to mention that much of the ensuing discussion of peasants' protests is based on official writings rather than any direct testimony from the peasants. In using officials' writings to reconstruct the nature of subaltern opposition, the strategy I follow is not very different from that used by a number of historians of colonial India (Spivak 1988: 203).

32 The annual forest report for 1921–22 explains the sharp rise in fires as follows: "The considerable increase in the number of fire cases is . . . mainly due to the terrible incendiarism which marked the hot weather of 1921, especially in the Kumaun circle, where the number of cases rose from 134 to 539, of which 331 were undetected. The number of fire cases taken into court rose from 16 to 139 and convictions were obtained in the great majority; the sentences were adequate running up to seven years' rigorous imprisonment" (GOUP 1922: 7). The report goes on to state that the decrease in the official number of unauthorized fellings is no guide to the real state of affairs.

33 Peluso and Vandergeest discuss the emergence and acceptance of the claims of the state to natural resources such as forests, parks, and wastelands (2002). Although their discussion concerns Southeast Asia, the processes they outline resemble those that occurred in Northwest India.

34 Similar activities by peasants were also noted in other parts of India, where there was organized defiance of the forest laws, sometimes as part of the Indian freedom struggle. During the Moplah rebellion in Madras, for example, many forest department buildings were looted and burned, records destroyed, and several forest guards assaulted (Anonymous 1923a: 327).

35 There was another wave of protests in Kumaon in the 1930s, which coincided with the Civil Disobedience movement launched by the Congress Party. Villagers set a large number of incendiary fires and on occasion threatened forest department employees with physical intimidation and violence (Robertson 1936).

36 Initially, the committee had three members: The district commissioner of Kumaon, the member of the legislative council from Garhwal, and a conservator from the forest service. An additional member, the chairman of the municipal board from Almora, was later appointed as a representative of the region (KFGC 1921).

37 The committee observed in its report that "the hill man is impatient of control, and we have it on record from the Deputy Commissioner and sub-divisional officers that any attempt to strictly enforce these [rules] would lead to riot and bloodshed" (ibid.: 1921, 3).

38 The report further argued that "the reservation benefitted Kumaon as a whole, there can be no doubt. When reservation of forests is first brought into effect, certain hardships naturally follow. This was found to be the case throughout India in the early days of the Forest department, but looking back at the results who can say that reservation was not the correct policy?" (Anonymous 1923b: 253).

39 This was the opinion expressed in the annual progress report of the forest administration for the province. According to the report, the new system was a failure because the "difficulties of the villagers are unsolved by any temporary relief obtained by wholesale destruction of the forests, and their later end will be worse than the first" (Anonymous 1927: 594).

40 As one Indian Forest Service officer remarked: "In West Almora it is essential to proceed with the utmost caution and forethought in all schemes to avoid clashing with the people's interests. In the political arena, the forests take a prominent place. . . . In the manufacture of forest grievances, and preparation of petitions to appellate authorities, the people are experts" (Turner 1929: 584).

41   The annual forest administration reports of the Madras Forest Department in this period often contain information on the transfer of local forests to panchayat management (Anonymous 1928: 462). The process of creating village forests to supply firewood for local residents had progressed by 1929 to a point where the Madras department had handed over nearly 20 percent of all its forests to forest panchayats. The actual area of community-managed forests in Madras thus amounted to 3,400 square miles, leaving 15,500 under the control of the forest department (Anonymous 1930: 356).

Part II A New Technology

1   With a modest beginning of 273 forest councils in the late 1930s (Robertson 1942: 120), the number of councils grew to nearly 3,000 by the end of 1995 (Agrawal and Yadama 1997).

2   For an iconoclastic analysis of the idea of empowerment currently popular in policy circles, see Cruikshank 1999.

3   For Foucault, the emblem of the coercive institution is Bentham's panopticon. It relies on a direct relation of visibility between power and the criminal. For a discussion, see the conclusion to chapter 5.

4   In contrast to the coercive institution, the punitive city operates less on the basis of a direct relation of visibility between the criminal and the supervisor than on the activation of a mental relation between a crime and its punishment by making the relationship visible through public examples. See the conclusion to chapter 5.

5   See Berry's (1993: 24–40) arguments about the efficiencies that indirect rule permitted in African countries. Although indirect regulation of forests through communities is quite different from indirect rule of native authorities as it unfolded in Africa, the economic logic of the two arrangements is comparable.

6   See Scott's argument about the inauguration of a new game of political rationality as a result of colonial rule that the colonized are obliged to play if they are to be considered political. According to him, the new game depends "on the construction of a legally instituted space where legally defined subjects [can] exercise rights, however limited they were" (1999: 45). Note also that Scott's views about the shift that colonialism inaugurated are in some conflict with Foucault's position on the extent to which sovereignty as a form of power was replaced by governmentality ([1978] 1991).

7   What I describe as features of monitoring within communities are

treated by Taylor (1982: 26–30) as more general core characteristics of community, especially (1) common beliefs and values, (2) direct and many sided relations, and (3) reciprocity.

8 Ostrom's (1990: 94–100) discussion of graduated sanctions is relevant here.

9 See ibid.; Agrawal and Goyal 2001; and Gibson and Marks 1995.

10 As Scott points out in his work on postcolonial criticism: "In the colonial world, the problem of *modern* power turned on the politico-ethical project of producing subjects and governing their conduct. . . . The political problem of modern colonial power was therefore not merely to contain resistance and encourage accommodation but to seek to ensure that *both* could *only* be defined in relation to the categories and structures of modern political rationalities" (1999: 52, emphasis in original).

11 Certainly, scholars of environmental politics who are less influenced by Foucault have paid even less attention to the making of subjectivities (Peluso 1992; Peluso and Watts 2001). See also as examples Baland and Platteau 1996; Bromley 1992; Ostrom 1990; and Wade [1988] 1994. Some work on colonial governmentality has begun to recognize the importance of the making of subjects, but it mostly explores the discursive processes involved and seldom engages the practices through which subjects come into being. See Chakrabarty 2000a; Prakash 1999; and Scott 1999.

12 In some ways, this part of the argument can be seen as an effort to examine how bureaucratic, scientific mechanisms of regulation reshape human agency (for a contrastive view, see Scott 1990).

13 This is true even of Jon Elster, who, among those who have written from an institutionalist perspective, is perhaps the most sensitive to questions about changing selves and subjectivities (see Elster, Offe, and Preuss 1998: 247–70; and Elster 1999).

Chapter 4 Governmentalized Localities

1 For a close examination of this point, see the review of the community-based conservation literature in Agrawal 1997. See also the careful review of the incentives of different actors involved in conservation programs in Wells 1998.

2 In 1917–18, for example, the annual report of the forest department argued that the actual burden of regulations and grazing dues was quite light. According to the report: "The feature of the rules which graziers object to is the restriction placed on their wandering with their cattle

from block to block, but without this restriction proper regulation on grazing would be impossible, and as there is no question of inadequate supply of grazing the hardship is more imagined than real. The increase in the grazing rates does not appear to be considered a grievance and in fact the industry of professional grazier is so lucrative that it could afford to bear with impunity a much heavier tax than the present annual grazing fee of 8 annas per cow or bullock and Re. 1 per buffalo" (GOUP 1917: 8). This was in a year when the forest department extracted an annual fee of Rs. 123,000 from cattle owners.

3   Baumann (1995) arrives at a similar inference after a review of evidence on precolonial communal management of forest resources: "Self-sufficient reliance on common property as an input into small holder subsistence agriculture that is suggested in the populist versions of the past was not a pre-British historic reality. The scarcities that arose through restrictions on forest use, as well as population growth under British rule, actually led to active management of village forests for the first time in many villages, and *a change in local norms and values relating to forests*" (56, emphasis added).

4   *Lattha* means "stick," and *lattha panchayats* literally means "councils based on the power of the stick." The name refers to the power the local community can exercise over members. References to lattha panchayats are available in several studies, but few of these provide in-depth accounts of their origins or evidence that their origins predated British rule (Nagarkoti 1997; Somanathan 1990, 1991).

5   According to Baumann (1995: 70): "Colonial rule in the early phase, rather than destroying indigenous systems of forest management, created the circumstances within which it was necessary for people to conserve resources. . . . [An] increase in subsistence demands on the forest, as well as the beginnings of unregulated commercial exploitation of the forests, drastically decreased the forest area available. The widespread reporting of collective action to conserve forests on village commons in the late nineteenth century and early twentieth century seems to be connected to this decline."

6   Nearly all temples in Kumaon have deodar trees near them. Because deodar is not native to Kumaon but to the region west of the Alaknanda, it is likely most of the trees near temples had a sacred status and were planted (Atkinson [1882] 1973: vol. 1, 325).

7   Batten, Kumaon's commissioner after Traill, remarked on the fact that "large portions of wastelands, including whole ranges and their vast forests, have been included from olden times in the boundaries of adjacent villages [as] such a division has been found useful in giving separate tracts of pasture for the cattle of different villages" (1878: 124).

8   It is unclear whether contemporary studies of informal forest councils in Kumaon can be seen as generating information about lattha panchayats as they existed prior to the formation of the forest councils under the rules of 1931. As the forest department pointed out in 1934, although only about 150 forest councils had been formed under the official rules, "in certain parts of Kumaun there are now a large number of village forests, which have been formed without official assistance on the lines of [official] Panchayat forests and as a result of the example given of the benefits of such forests" (GOUP 1934: 1).

9   Anthropological work in the 1950s and 1960s was beginning to cast doubt on some of these views of Indian villages. But its very focus on village India helped consolidate the categories it sought to question (Lewis 1958; Marriott 1955; Srinivas 1960). Nor, it must be admitted, were writings during this period particularly careful about how they deployed categories such as village and caste.

10   Metcalfe's account is cited by Kessinger (1974: 25):

> The village communities are little republics, having nearly everything they want within themselves, and almost independent of any foreign relations. They seem to last where nothing else lasts. Dynasty after dynasty tumbles down; revolution succeeds to revolution; Hindoo, Patan, Mogul, Mahratta, Sik, English are all masters in turn; but the village communities remain same. . . . If a country remain for a series of years the scene of continued pillage and massacre, so that villages cannot be inhabited, the scattered villagers nevertheless return whenever the power of peaceable possession revives. A generation may pass away, but the succeeding generation will return. The sons will take the place of their fathers; the same site for the village, the same position for the houses, the same lands, will be occupied by the descendants of those who were driven out when the village was depopulated.

I came across this reference in Ludden 1993. I am indebted to Inden (1990) for part of the discussion that ensues.

11   One of the reasons that such visions of Indian villages became popular was how well they could be used to suggest that the "removal of the thin conquering strata of Europeans and the Pax Britannica enforced by them would open wide the life and death struggle of inimical castes and tribes" (Weber 1958: 325).

12   See, on this score, the criticisms offered by Katten (1999: 88–89) of scholars such as Washbrook and O'Hanlon (1992), Chatterjee (1993), and Prakash (1990), who have tended to take the ideas of nation and class for granted in their work. The problem, as he points out, is at least partly one of limited but convenient historical categories. But the end result is that

scholarship on India is left subject to the "set of meanings and ideologies accompanying these basic and insufficient categories," such as nation, class, and village (Katten 1999: 89). Ranajit Guha's arguments about "optics" (1983) and Chakrabarty's observations about "knowledge protocols of academic history" (1992: 23) point toward similar limits of analysis. See also Stoler's (1989) valuable discussion of colonial analytical and political categories.

13   Even as the forest department was preparing plans to reclassify nearly all the nonagricultural land in Kumaon as reserved forest, it recognized that "the allowances made for right-holders [villagers] appear likely to prove so inadequate that the scheme must be completely revised when the schedules of rights are received from the Forest Settlement Officer" (GOUP 1915: 9).

14   Forest department officials so disliked the recommendations proposed by the Kumaon Forest Grievances Committee that they even welcomed the formation of a permanent forest advisory committee in 1922. They argued that the remedies advanced by the committee forced the department to release large areas of forests "before any alternative system had been developed . . . [and] to meet this defect the committee was considering measures for giving practical and general effect to the proposal put forward many years ago to form communal forests" (GOUP 1925: 2).

15   The remaining provisions, which are discussed at length in the next chapter, were mostly guidelines about the internal functioning of the councils.

16   This last group of rules allowed the councils significant latitude in the way they governed, given the externally prescribed limits on council authority (Uttar Pradesh Government 1931).

17   The 1931 rules also underwent some changes in 1972, but these were relatively minor compared to what happened in 1976.

18   I defer until the next chapter the discussion of rules that impinge on the relationships between the new decision-making units in the community and common village residents.

19   Rules listed under the heading of "Membership, Officials, and Decision-Making" were mainly coordination rules designed to facilitate the functioning of the councils, and I do not discuss them in any detail. But it is interesting that the modifications of 1976 made it more difficult to reach any decision, as a two-thirds majority was now required. I am not sure why this change was made, but it is unlikely that it made much difference. In the records of more than thirty-eight councils that I examined (on the average each council had records for at least ten years), most decisions of the governing committee were unanimous.

20 Author's interview 2 with Shankar Ram, tape 1, translated by Kiran Asher.

21 For a discussion of visibility and legibility, see Scott 1998.

22 Forest department officials explained in 1926 that, although nearly 2,000 square miles of class I forests had been transferred to the district revenue authorities from the control of the forest department on the recommendations of the Kumaon Forest Grievances Committee, they were "mostly oak and miscellaneous forests many of them covering hills the stability of which is of prime importance" (GOUP 1926: 1).

23 The figures on the number of forest councils at mid-century are to be found in UPFD 1959, 1961. For more recent figures, I rely on fieldwork and data collection from 1995.

24 See Schelling 1980 for a lucid discussion of the idea of focal points.

25 Braddick notes the value of precision in the exercise of decentralized authority in his fine study of state formation in seventeenth century England: "By reducing the discretion available to specialized state functionaries, and by specifying their rewards more closely, the emphasis on precision and regularity helped to secure more ready consent" (2000: 43).

26 As early as the 1930s, the forest department was observing efforts to form forest councils. According to one report, the forest council movement "still requires official guidance to obtain the best results by good organization, but it appears already to have become an established feature and one of the most remarkable developments of recent times in Kumaun" (GOUP 1934: 1).

27 I do not enter into a discussion of the vast literature that engages the issue of state autonomy and state-society relations. Some representative accounts are available in Evans, Rueschemeyer, and Skocpol 1985; Hall 1986; Jessop 1990; Migdal 1988; and Migdal, Kohli, and Shue 1994. A thought-provoking and useful critique of the statecentric literature is available in Mitchell 1991.

28 The extension of rules into state formations, or the incorporation of territories through such rules, is not necessarily a threat. But see Brow 1996, which argues that development in Sri Lanka incorporates villages into regional and national circuits of power and exchange: "Various social practices that had served to mark the inhabitants of the same village as members of a distinct community were under threat while others had already been abandoned" (6).

29 Ferguson goes on to say that the growth of state power means the state "grabs onto and loops around existing power relations, not to rationalize or coordinate them, as much as to cinch them all together into a knot . . .

[and] it is involved in the distribution, multiplication, and intensification of these tangles and clots of power" ([1990] 1994: 274). Although it is not clear what exactly "knots, tangles, and clots of power" means, it does seem that in saying that "the state grabs onto" or "is involved" Ferguson is thinking of the state as a singular, unitary actor.

Chapter 5 Inside the Regulatory Community

1 See McKean's (2000: 49) review of factors that are likely to promote successful common property regimes. She argues for building on existing institutions rather than trying to craft entirely new ones. Ostrom, Schroeder, and Wynne make a similar point in relation to development processes when they suggest that indigenous institutions can serve "as an important foundation upon which to construct a social infrastructure that [is] consistent with a modern democratic political economy" (1993: 7).

2 William Ophuls argued, for example, that the problems of environmental commons that Garrett Hardin (1968) discussed were a special case of the general political dynamic of Hobbes's state of nature. According to Ophuls, the political dilemma of ecological scarcity was "authoritative rule or ecological ruin" (1973: 162). For a review of some of the evidence on this subject see Ostrom 1990; and Ostrom, Walker, and Gardner 1992: 404–5.

3 Chapter 1 mentioned the ubiquity of community-based programs aimed at protecting nature. See the discussion of more than fifty-five such programs in thirty countries in Africa and Latin America in Agrawal 2001b. See also Agrawal and Gibson 2001; Bryant and Bailey 1997; Clay 1988; Ellen, Parkes, and Bicker 2000; Gibson, McKean, and Ostrom 2000; Peluso 1992; Poffenberger and McGean 1996; Western and Wright 1994; and Wolverkamp 1999. Many of these books and collections of studies present examples of the importance community and decentralized regulation have assumed in environmental conservation.

4 See, for example, Ehrlich and Ehrlich 1991; Meadows, Meadows, and Randers 1992; Meffe, Ehrlich, and Ehrenfeld 1993; Myers 1991; and Wilson 1992. Indeed, concern with overpopulation has colonized even seemingly unrelated subjects: Lévi-Strauss summarizing his views on race and culture (1985: 21), Jack Nicholson talking about solar energy (1991: 165), or Crick and Watson explaining their discovery of the double helix structure of DNA (Jaroff 1993: 59).

5 As one might suspect, they draw much of their inspiration from the well-known work of Malthus ([1798, 1803] 1960). For some examples of mod-

ern day neo-Malthusian writings, see Avise 1994; Durning 1989; Fischer 1993; Hardin 1993; Holdren 1992; Low and Heinen 1993; Ness, Drake, and Brechin 1993; and Pimental et al. 1994.

6 See Carrier 1987; and Stocks 1987: 119–20. See also the arguments in Arizpe and Velazquez 1994: 23.

7 See Chomitz 1995; Fearnside 1986; Verma and Partap 1992; and Young 1994.

8 In viewing regulations as institutional forms, I am drawing on the definition of institutions proposed by Bates (1989) and North (1990).

9 Scholars of common property have played a significant role in bringing community-based conservation to the fore. Ostrom 1990 remains the central text in the field of common property. Baland and Platteau 1996 and Wade [1988] 1994 constitute important supplementary texts. Agrawal 2001c provides a useful summary and critical review of the literature. Political ecologists have also contributed a steady stream of writings that favor an important role for local populations in resource management. See Neumann 1992, 1998; and Peluso 1992.

10 This literature focuses primarily on relatively autonomous community-level governments or at least on forms of community-based government that are not directly and formally affiliated with state actors. It came into prominence with the publication of a collection of essays by the National Research Council in the United States (National Research Council 1986) and Ostrom's seminal work on the commons (1990). A large number of essay collections and significant empirical studies spanning the globe have helped the scholars of commons make their point convincingly: community-level governance of resources can be as effective as government through the state or market-based mechanisms (see Acheson 2002; Berkes 1989; McCay and Acheson 1987; McKean 1992; Ostrom, Gardner, and Walker 1994; Peters 1994; Pinkerton 1989; and Sengupta 1991).

11 The full analysis is available in Agrawal and Yadama 1997. The 1997 essay presents descriptive statistics, details on sampling and data collection, information about the model we used, procedures for estimating specific variables, and the numerical estimates of various relationships. Since the main point of the argument in this chapter is simply to indicate the importance of local regulation, I report the findings of the previous study rather than undertaking a full-blown repetition of the analysis.

12 The use of fire to clear vegetation for agricultural purposes was widespread in many parts of British India, most notably in Burma, the Deccan, the Central Provinces, and the Northeast, and it attracted enormous criticism and restrictions from forest department officials (Stebbing 1922–26). But in Kumaon the practice was uncommon at best by the early twentieth century.

13   My discussion of the effects of the Forest Council Rules is based on fieldwork conducted during visits to thirty villages between 1989 and 1993. A research team helped collect the bulk of the data in the summer of 1993. I met several members of the research team and discussed some of our findings during a final visit in the spring of 1997.

14   Taylor remarks, in his interpretation of Wittgenstein's (1973) work, that "following rules is a *social* practice" (1993: 48, emphasis in original). See also Kripke 1972 for an alternative interpretation of Wittgenstein that emphasizes the difficulty of reasonably justifying why particular implications of rules are followed. But even these alternative interpretations accept that rules in themselves do not contain all of their possible meanings. Rule-following behavior accepts particular implications of rulelike statements even if they are not ultimately defensible.

15   "Right holder" is the literal translation of the local term *haqdar*. Typically, all households belonging to a settlement hold formally equal rights to harvest forest products from the council-managed forests. The amount of produce that can be harvested by each right-holding household is fixed by the forest council. The term signifies the contractual nature of the relationship between the forest councils and village households.

16   Elinor Ostrom, personal communication, December 10, 2000.

17   Such sentiments stand in sharp contrast, of course, to other expressed idioms in which villagers indicate that everyone knows what happens in a village. The point is that it is difficult to catch someone in the act when he or she is removing fodder or firewood, even if there is a general awareness of how specific individuals act. In other studies of village life or life in small groups, it is a commonplace that members know a great deal about those with whom they interact frequently (Agrawal 1997). But general awareness of others' reputations does not mean widespread knowledge of their daily actions.

18   See Elster 1989: 40–41; and Ostrom, Gardner, and Walker 1994: chap. 1.

19   See Fearon (1999), who suggests that at election time voters are more likely to favor candidates they see as good performers than to sanction poorly performing officeholders. The distinction dissolves when voters perceive a strong link between past decisions of officeholders and their likely future performance *or* when candidate behavior is widely known among voters, as is likely in the small settlements where forest council officials are elected.

20   Chapter 6 explores and explains why it is that some villagers conform to new forms of government through the community and others withstand the production of conformity. In both cases, changes in subjectivity occur as strategies of government.

21 Much of this basic information about Bhagartola is available in the records of the *patwari*, the local revenue department official.

22 Women' subordination is thus not simply the result of policies imposed by the state. For excellent analyses of women's subordination through different strategies of control over property, see Agarwal 1992, 1994.

23 In any case, the more important question is not whether power is exercised unequally now and was more equal in a distant past, when the government of environment was less influenced by powerful state officials. The more pressing question is *how* unequal power unfolds today compared to how it was exercised prior to the passage of the Forest Council Rules of 1931. The introduction of new institutions has made the exercise of unequal power more subtle. One aspect of this change is that politics can no longer be seen (if it ever was) to be a characteristic only of the relationships between the state and the community.

24 Of course, in his later work on governmentality Foucault arrives at very different conclusions about the extent to which modern power disindividualizes or is homogeneous in its effects on humans ([1978] 1991).

Chapter 6 Making Environmental Subjects

1 I owe the expression "boundary work" to Donald Moore and Tania Li (personal communication, 1998).

2 Anderson borrows the term from Seton-Watson but gives it a bite all his own ([1983] 1991: 86).

3 It is precisely to this politics that Chakrabarty (2000a), indebted no doubt in important ways to Chatterjee (1986, 1993), draws attention when he seeks to "make visible the heterogeneous practices of seeing" that often go under the name of imagination. Chakrabarty carefully traces the differences among the many ways of imagining the nation by examining peasants and a literate middle class.

4 Inattention to this politics in Anderson's account is signaled, of course, at the very beginning of his cultural analysis of nationalism. After defining *nation* as "an imagined political community—and imagined as both inherently limited and sovereign" ([1983] 1991: 6, 7), he intimately examines each term in the definition—except *political*. It is not only Anderson's history of nationalism that can be enriched by attending to the politics of subjecthood but also his view of culture more generally.

5 The essays in Guha and Spivak 1988 constitute one of the best introductory texts about subaltern studies. See Guha 1982b, 1997; and Chatterjee and Jeganathan 2000 for a sense of the different moments in the life of a

collective. The collection of papers in Ludden 2001 constitutes a fine example of some of the more careful critical engagements with the work of subaltern studies authors.

6   Guha 1982a: 4–6. For a more recent consideration, see Guha 1997.

7   At the same time, it is fair to observe that more recent scholarship in a subalternist mode has begun to take Foucault's ideas about power and subject formation more seriously and has carefully examined how different kinds of subjects came into being both under colonialism and in modernity (Arnold 1993; Chakrabarty 2000b; Prakash 2000).

8   By total institutions, I mean those Foucault refers to as "complete and austere institutions" ([1975] 1979: 233); prisons, concentration camps, and insane asylums are prime examples.

9   During my first visit, I talked with a total of forty-three villagers. I could not meet and talk with eight of them in 1993 for a variety of reasons. Some had moved out of the village, some could not be located, and one had died.

10   The inability of the state to protect property in the face of concerted resistance is of course not a feature of peasant collective action in Kumaon alone. The threat to established relations of use and livelihood that the new regulations posed is similar to the kinds of threats that new technologies and institutions have posed in other regions as well. For example, the invention of mechanized implements has often sparked such responses from peasants and agrarian labor; these have found some success precisely because of the inability of the government to detect them (Street 1998: 587).

11   I have reported statements and actions by various persons in two different time periods as being representative of the groups to which they belong, a common strategy for scholars belonging to fields as different in their assumptions as cultural anthropologists and rational choice political scientists. Bates 1981, 1989; and Bates, Figueiredo, and Weingast 1998 are rational choice exemplars of this strategy. Ferguson [1990] 1994; and Gupta 1998 are counterpart examples from cultural anthropology.

12   Author's interview 2, with Shankar Ram, tape 1, translated by Kiran Asher.

13   Author's interview 13, with Bachi Singh, tape 5, translated by Kiran Asher.

14   In Indian mythology, Kaljug is the fourth and the final era before time resumes and proceeds through the same sequence of eras: Satjug, Ttreta, Dwapar, and Kaljug. This is the time when dharma gives way to *adharma* and established authority fails.

15   Miller and Rose (1990) follow this argument closely in elaborating the concept of "government at a distance" and examining how modern gov-

ernment overcomes the natural diluting effects of distance in the exercise of power.

16 In coining the phrase "intimate government" I would like to acknowledge a debt to Hugh Raffles (2002), who uses the idea of intimate knowledge in talking about indigenous knowledges and their circulation in the corridors of policy making.

17 Much of the literature on environmental politics that uses an analytic of domination/power and resistance/marginality remains encoded within this structural division between freedom and constraint as well. See, for example, Brosius 1997; and Fairhead and Leach 2000. More general studies of domination/resistance are subject to the same tendency (see Kaplan and Kelly 1994; and Lichbach 1998).

18 See Latour 1999 (15, 294) for a thought-provoking examination of scientific practice.

19 Rorty (1984) complains that Foucault is a cynical observer of the current social order rather than one to whom that order is important. Dews (1984), calling Foucault a Nietzschean naturalist, asserts that his insights cannot be a substitute for the normative foundations of political critique. According to Fraser (1989: 33), Foucault adopts a concept of power that "permits him no condemnation of any objectionable feature of modern societies . . . [and] his rhetoric betrays the conviction that modern societies are utterly without redeeming features." Taylor (1984) advances perhaps the strongest argument in this vein, suggesting that Foucault's account of the modern world as a series of hermetically sealed, monolithic truth regimes is as far from reality as the blandest whig perspective of progress. See also Philp 1983. For close and persuasive arguments that engage these critiques of Foucault's ethics and go a long way toward showing their logical and interpretive gaps, see Dreyfus and Rabinow 1983; and especially Patton 1989.

20 Butler also emphasizes the linguistic and psychic aspects of the constitution of the subject. Given my interest in locating the social mechanisms through which subjects come into being, a focus on psyche and language would lead us far astray. As Rose ([1989] 1999: xix) argues, language is only one element in the way one's relationship to oneself is shaped and reshaped historically.

Chapter 7 Conclusion

1 As Al Gore puts it: "We must make the rescue of the environment the central organizing principle for civilization" (1992: 269). Buell's (1995: 2)

reflection on the accuracy of this statement closely matches the perceptions of many environmentalists—"no informed person would contest
that it expresses an anxiety much stronger today than ever before in
recorded history, and likely to grow stronger."

2   See Ribot 1995.

3   See White 1995: x.

4   For a comprehensive review of empirical studies of local management of
commons, see Baland and Platteau 1996. Region- and country-specific
discussions and case studies of local management are numerous and
widely available. For some exemplary studies, see Berkes 1989; Berkes
and Folke 1998; Fernandes, Menon, and Viegas 1988; Gibson, McKean,
and Ostrom 2000; McCarthy et al. 1999; McCay and Acheson 1987;
McKean 1992; National Research Council 1986; Ostrom 1990; Peters
1994; Pinkerton and Weinstein 1995; Poffenberger 1990; Redford and
Padoch 1992; and Wade [1988] 1994.

5   Agrawal 2001b examines fifty-five cases of environmental policy changes
in Africa and Latin America in which communities and local populations
have come to play some decision-making or implementation role.

6   The failure has occurred along a number of dimensions: highly unequal
distributive effects, resistance from those considered marginal and powerless, and steady erosion of land under vegetation (Agrawal 1997; CSE
1982). Agrawal (1998: 59) argues that even the early efforts of state agencies to address some of the deficiencies of centralized control faced
"widespread local resistance, including people uprooting saplings" that
forest department officials had planted.

7   Several other distinct streams of environmental writings have identifiable disciplinary origins: environmental anthropology, historical ecology,
human geography, environmental history, and environmental sociology.
Insights from disciplinary environmentalist scholarship have significantly influenced the more cross-disciplinary literatures on common
property, political ecology, and ecofeminism that are the direct focus on
this chapter. I have chosen to focus on more cross-disciplinary studies in
part because of their relatively explicit attention to policy and politics. In
addition, I am concerned in this chapter mainly to flesh out a robust
understanding of environmental politics by tracing the historical articulation of power with nature. Given the orientation of this chapter,
disciplinary crossfire is clearly less interesting.

8   Although I choose the work on common property as the entry point for
my discussion of these three interdisciplinary streams of literature, I do
not mean to privilege any ontological or chronological priority by my
choice. The roots of political ecology can be traced back to discussions in

political economy and cultural ecology, and the ancestry of ecofeminism can be handily discerned in phenomenology and early feminist and environmentalist writings. For example, Sturgeon sees in ecofeminism a name that "usefully if partially describe[s] the work of Donna Haraway and Mary Daly, Alice Walker and Rachel Carson, Starhawk and Vandana Shiva" (1997: 24).

9   A number of scholars associated with Ostrom have helped extend the research program on the effectiveness of community-level institutions in the management of renewable resources (see Blomquist and Ostrom 1985; Lam 1998; Ostrom, Gardner, and Walker 1994; Schlager and Ostrom 1992; and Tang 1992).

10   If common property theorists have begun to pay attention to politics, albeit through the mechanism of institutional reconfiguration, they have yet to initiate any serious examination of the relationship between institutions and the production of environmental identities or the connections between changes in institutional arrangements and transformations in the character of environmental subjects. I examine this gap below in summing up the contributions of these three sets of interdisciplinary writings on environmental politics.

11   Although Peet and Watts (1996: 2) identify political ecology as one factor worth citing (together with the collapse of socialism and the resurgence of global environmental concerns) to explain a greater emphasis on nature-society relations, it may be appropriate to claim a somewhat more modest role for interdisciplinary analyses of the environment, including those by political ecologists.

12   Initial works in political ecology can be dated perhaps to the 1970s, to papers presented in a symposium on nature and environment that were published in *Anthropological Quarterly* and especially to a short paper by Eric Wolf entitled "Ownership and Political Ecology" (1972). Some of the points in Wolf's paper were developed at greater length by Cole and Wolf ([1974] 1999). But the substance of the arguments advanced in the 1970s in Wolf's paper, despite its title, is at variance with later theoretical innovations in this field in several respects. For example, despite his emphasis on processes and politics behind rules, Wolf also retains a belief in "self-regulating communities" in parts of the Alps in contrast to another ideal-typical formation—politically federated groups in the valleys. Nor does he say much about what political ecology might be. A clearer ancestry might be traced to the late 1970s and the early 1980s (Blaikie 1985; Blaikie and Brookfield 1987; Bunker 1985; Cockburn and Ridgeway 1979; Redclift 1984; Watts 1983).

13   The notable exception to the Malthusian focus of early environmentalist

writings is Carson (1962), who was far more concerned with the effects of artificially produced chemical compounds and the role of corporate actors in ignoring the effects of the chemicals they sold as an unmixed good.

14 For concise reviews of the multiplicity in political-ecological approaches, see Peet and Watts 1996: 6–13; and Bryant and Bailey 1997: 20–26. After pointing to the limited and undertheorized role of politics in Blaikie and Brookfield's work, Peet and Watts list several new directions for political ecology: to theorize the specific dynamics of environmental change and socialism; to attend to resistance, social movements, and organized politics as examples of politics related to the environment; to relate civil society to environmental associations and organizations; to analyze environmental discourses and narratives; to deepen the historical aspects of environmental change; and to shift the analysis of ecological processes from idioms of harmony and stability toward those of complexity, chaos, and disequilibrium. The grab bag of approaches that they list does not improve when it is compared to that of Bryant and Bailey. Among the approaches they identify are those oriented toward addressing specific problems (soil erosion, deforestation, overfishing), concepts (social construction of the environment, scientific forestry, sustainability), regions, social categories (such as class, gender, or ethnicity), or actors (peasants, corporations, nongovernment organizations). As a way to bring greater unity to this wide diversity, Peet and Watts look to and advocate poststructuralism for theoretical inspiration and social movements for inspiration in social organization. Bryant and Bailey, focusing on the existing gaps in political theorizing in political ecology, advocate an actor-oriented approach.

15 Peet and Watts (1996: 263) attempt to exercise a significant amount of care in how they see the social as being naturally constructed and in their advocacy of specific elements of poststructuralist thought. But, no matter how much they wish it, a natural construction of the social would require some presocial, prediscursive entity that can be called "natural."

16 See Neumann 1992; and the special issue of the journal *Antipode* edited by Neumann and Schroeder (1995).

17 A range or writings that can be identified as ecofeminist began to flower in the late 1970s and blossomed from the 1980s onward (Diamond and Orenstein 1988; Merchant 1980; Warren 1987, 1997). I would like to acknowledge Bina Agarwal's influence on several of this section's arguments about gender-environment relationships and Donald Moore's helpful suggestions for a title.

18 Kathy Ferguson's analysis of three kinds of essentialisms in relation to feminism, and the tendency among different theorists to conflate them,

is worth examining (1993: 81–90). According to Ferguson, the first, essentialism per se, refers to arguments that attribute women's experiences to some unchanging traits in physiology or some larger order of things. The second, universalism, takes the patterns visible in a particular time and place as being accurate generally. The third, the constitution of unified categories, entails the creation of any unified set of categories around the terms *woman* and *man*. Ferguson suggests that the third kind—the constitution of unified categories—is fundamental to language and that all analysis requires such naming practices. Although it is possible to be mindful of the contingent and specific nature of the creation of such categories, it is impossible to operate without them. Often charges of essentialism confuse these three forms. Such confusion is especially problematic when instances of the third kind are criticized as being of the universalist or biological kind.

19   Some of the differences prompting scholars to claim different names or typologies for their approaches are a function of the emphasis they place on the difference between anthropocentrism/androcentrism versus ecocentrism, and environment versus nature/ecology. See Eckersley 1992 for a discussion of these differences and Grendstad and Wollebaek 1998 for an empirical study. For an examination of the typologizing processes that produce feminisms with different names and characteristics, see King 1994. Her argument applies equally to feminist environmentalisms. The typologizing impulse and charges of essentialism often have the unfortunate consequence of eliding the political contexts within which particular forms of feminist environmentalism may have come into being (Sturgeon 1999).

20   Among the most important of the conditions noted by Warren are (1) oppression of women and nature shares important connections; (2) the nature of these connections is key to understanding the oppression of women and nature; and (3) feminist theory must include an ecological perspective and solutions to environmental problems must include a feminist perspective (1987: 4–6).

21   In this regard, see also McMahon 1997, which launches a critique of neoclassical economics from an ecofeminist stance.

22   Foucault's lecture on governmentality at the College de France was first published in the journal *I & C* in 1979. It has come to be more widely known in its revised form in Burchell, Gordon, and Miller 1991.

23   Foucault's definition of *government* as the "conduct of conduct," or the effort to shape, guide, or affect the conduct of some agent(s), underpins the principal sense in which the term is used throughout the following discussion.

24   See Burchell, Gordon, and Miller 1991: ix. In his lectures, Foucault himself applied this perspective to a number of historical domains: the idea
of government as a form of "pastoral power" in Greek antiquity, doc-
trines of government associated with the reason of state and police in
early modern Europe, the eighteenth-century beginnings of liberalism,
and postwar forms of neoliberal thought in Germany, the United States,
and France. Dean (1999: 3–4) lists some of the specific works that describe and examine how modern forms of power relate to politics in
these different domains.

25   Foucault speaks explicitly of the emergence of biopower as taking place
in the seventeenth century and treats it as a new form of power over life.
His discussion of governmentality (Foucault [1978] 1991) can be seen as
an effort to track changes in forms of political control as they relate to
the triangle of sovereignty, discipline, and government. Two developments underpinned the appropriation by states of power over life: the
shattering of feudal structures and the establishment of large territorial,
administrative, and colonial states; and the Reformation and Counter-
Reformation, which raised the question of how one must be ruled spir-
itually.

26   Power over life can be seen to operate in two basic forms. Disciplinary
power is centered on the body, optimizing its capacities, increasing its
usefulness and docility, and integrating it into systems of efficient and
economic controls. The other form is centered on the entire species and
aims at the regulation of the biological processes that affect a whole
population: birth, mortality, health, and life expectancy. Institutions
such as the army, schools, barracks, and workshops came to embody the
mechanisms used to discipline the body, whereas the regulation of population was achieved by techniques in emerging fields such as demography, economics, statistics, and resource management. In these fields, we
witness the methods and techniques that can optimize life and its forces
without also making it more difficult to govern. The careful shepherding
of economic processes and the forces that sustained them became a
rationale for governance (Foucault 1978: 139–41).

27   "Omnes et Singulatim" is the title of Foucault's Tanner Lecture, which
was delivered at the University of California, Berkeley (Foucault [1979]
2000).

28   This style of political governance was fully articulated for the first time,
according to Foucault, under the rubric of *polizeiwissenschaft*, the science
of police. The idea of policing, with the appropriate translation of *polizei*
being closer to "policy" than to "police," developed in German territories after the Thirty Years War.

29 Gordon (1991: 10–11) draws on Foucault's lectures at the College de France in 1978 to make these comments. See Pasquino 1991 for a further discussion of this particular relationship between the ideas of police and policy.

30 Indeed, the implicit assumption that the individual is the link between states and societies is precisely the reason why those who try to undermine the state-society separation attempt to show the simultaneous location of state officials in society and the links of individual members of society with the state (Gupta 1995; Migdal, Kohli, and Shue 1994; Nugent 1994).

31 Hacking notes in his important historical study that "statistics has helped determine the form of laws about society and the character of social facts. It has engendered concepts and classifications within the human sciences. . . . It may think of itself as providing only information, but it is itself part of the technology of power in a modern state" ([1981] 1991: 181).

32 The chief virtue of liberalism, according to Foucault, is that it grants due credit to those regular and natural processes that characterize aggregates such as the population or the economy. Liberal government does not try to impose on them the will of a sovereign. Internal regularities of aggregate phenomena become evident because of the application of specific procedures of knowledge generation and expert authority creation. The importance of new knowledges lies in the fact that they show us how to govern, how to regulate, and how to achieve desired ends. The difference between the state that tries to realize the full import of sovereignty and the state that attempts to govern is precisely that the one imposes a sovereign will whereas the other seeks to deploy the correct forms of surveillance, control, and management in order to achieve optimal outcomes. It is precisely to highlight this difference between liberal governmentality and earlier forms of state power that Foucault pays so much attention to the ideas of pastoral power and likens government to the art of ensuring the "correct disposition to things, arranged so as to lead to a convenient end" ([1978] 1991: 93).

33 Indeed, the influential collection of essays on governmentality edited by Burchell, Gordon, and Miller represents itself as a "genealogy of the welfare state—and of neo-liberalism" (1991: 37). Another recent volume on governmentality focuses on its liberal forms (Barry, Osborne, and Rose 1996). Similarly, a larger proportion of the essays related to Foucault's work in the journal *Economy and Society* takes governmentality as a feature of power in modernity. See also Grant's (1997) interesting essay on discipline and its relationship to the formation of subjectivities in the context of schooling.

34 Foucault was to remark in the later period of his life that "the goal of my work during the last twenty years . . . has not been to analyze the phenomena of power. . . . My objective, instead, has been to create a history of the different modes by which, in our culture, human beings are made subjects" (1982: 208).

35 See Connolly 1985; and Patton 1989 for a defense of Foucault's views on power and for an explication of how these views remain consistent across his early and later writings.

36 In its primary sense, power for Foucault is "power to," the ability of a subject to accomplish something, and it is exercised whenever there is action upon the actions of others. Relations of power can exist only when they involve forms of action upon the actions of others and leave open a range of possible responses (Patton 1989: 271). Thus, for Foucault freedom and power do not exist in an oppositional relationship. There is less face-to-face confrontation between power and freedom and more permanent provocation.

37 Remarking on this view of power, Taylor says that "the utter sterility of the view popular a while ago in American political science, that one could analyze power in terms of A's ability to make B do something he otherwise would not illustrates this [Foucault's invaluable contribution]. . . . Acts of power are so heterogeneous; they absolutely do not admit of being described in such a homogeneous medium of culturally neutral makings and doings" (1984: 171). One can add to Taylor's scathing criticism. Views of power in which A makes B do something without B realizing it continue to adhere to a negative, sterile view of power.

38 Positive freedom, in Isaiah Berlin's discussion, refers to desires for self-government and autonomy (1969). But for Foucault the existence of individual capacities underlies the idea of positive (and negative) freedom (Patton 1989: 262). Internal constraints, which depend on the intellectual and moral constitution of a person, can limit the class of actions that a person is able to perform as much as external constraints do. Positive freedom, then, refers to the degree to which internal constraints limit a person's ability to undertake some action.

39 See, for example, Foucault's own investigations of subject formation in periods outside of modernity and through techniques that broadly fall under what one might call the "cultivation of the self" ([1985] 1990, [1986] 1988).

40 See Agrawal 1995 for a review.

41 Recall in this context the strongly resonant beginning of Foucault's *The Order of Things*, where he cites Borges on a "certain Chinese encyclopedia" that presents an "unthinkable" classification of animals (1970: xv–

xvii). The taxonomy from a given system of thought demonstrates for Foucault the stark impossibility of being imagined if one inhabits a different system of thought. Under modernity, perhaps, a particular variant of reason—narrow rationality—has come to constitute the entirety of what could be considered reason.

42 Foucault called his lectures on the subject "The Government of One's Self and of Others" (Gordon 1991: 2).

43 Ferguson's and Gupta's use of Foucault can be seen to conform to what Hannah calls the evocative rather than the exegetical (2000: 4). Gupta's analysis is discussed below.

44 Later in the book, Ferguson defines *governmentality* as "the idea that societies, economies, and government bureaucracies respond in a more or less reflexive, straightforward way to policies and plans. In this conception, the state apparatus is seen as a neutral instrument for implementing plans, while the government itself tends to appear as a machine for providing social services and engineering economic growth" ([1990] 1994: 194). Nothing that Foucault says about governmentality, however, signals that the term refers to the neutrality of the state or that it simply denotes the expansion of modern states. A reading of Foucault through an organizational theoretic optic is likely to recognize only the increasing efforts by states to produce regulation. Such a reading misses some of the most provocative and interesting elements in the concept of governmentality, reducing Foucault at best to a version of Weber.

45 Two other important recent efforts to analyze development processes (and to some extent issues related to environmental conservation as well) from a Foucauldian stance are those of Escobar (1995) and Gupta (1998). Escobar's critique of strategies of development since the 1950s dwells more on the concepts of biopower and biopolitics than on governmentality and does not examine questions of changes in subjectivities or identities. His work attends more to the manner in which discourses structure a certain view of development among both practitioners and those who have been its victims. Gupta suggests the possibility of a "new regime of discipline in which governmentality is unhitched from the nation state to be instituted anew on a global scale" (321). For him, governmentality is about the "global regulation of populations, bodies, and things" (34) as a result of new global treaties and accords. Global institutions construct a particularly constrained field of beliefs and actions for third world peasants and thereby inaugurate a form of domination different from that of the nation-state. In both of these accounts the use of Foucault is aimed at a particularly dichotomized conceptualization of power and its effects.

Bibliography

Abrams, Philip. 1988. Notes on the difficulty of studying the state. *Journal of Historical Sociology* 1:58–89.

Acheson, James. 2003. *Capturing the Commons: Devising Institutions to Manage the Maine Lobster Industry.* Hanover: University Press of New England.

Agarwal, Bina. 1992. The gender and environment debate: Lessons from India. *Feminist Studies* 18:119–58.

———. 1994. *A Field of One's Own: Gender and Land Rights in South Asia.* Cambridge: Cambridge University Press.

———. 1997. Gender, environment, and poverty interlinks: Regional variations and temporal shifts in rural India, 1971–1991. *World Development* 25(1): 23–52.

———. 1998. Environmental management, equity and ecofeminism: Debating India's experience." *Journal of Peasant Studies* 25(4): 55–95.

Agrawal, Arun. 1994. Rules, rule-making and rule-breaking. In *Rules and Games*, edited by E. Ostrom, R. Gardner, and J. Walker, 267–82. East Lansing: University of Michigan Press.

———. 1995. Dismantling the divide between indigenous and western knowledge. *Development and Change* 26(3): 413–39.

———. 1997. Community in conservation: Beyond enchantment and disenchantment. CDF Discussion Papers, no. 1, University of Florida, Gainesville.

———. 2000. Small is beautiful but is larger better? Forest management institutions in the Kumaon Himalaya, India. In *People and Forests: Communities, Institutions, and Governance*, edited by Clark C. Gibson, Margaret A. McKean, and Elinor Ostrom, 57–85. Cambridge, Mass.: MIT Press.

———. 2001a. State formation in community spaces? The forest councils of Kumaon. *Journal of Asian Studies* 60(1): 1–32.

———. 2001b. The decentralizing state: Nature and origins of changing environmental policies in Africa and Latin America, 1980–2000. Paper presented at the Ninety-seventh Annual Meeting of the American Political Science Association, San Francisco, August 30 to September 2, 2001.

———. 2001c. Common property institutions and sustainable governance of resources. *World Development* 29(10): 1649–72.

Agrawal, Arun, and Clark Gibson. 1999. Community and conservation: Beyond enchantment and disenchantment. *World Development* 27(4): 629–49.

Agrawal, Arun, and Clark C. Gibson, eds. 2001. *Communities and the Environ-*

ment: Ethnicity, Gender, and the State in Community-Based Conservation. New Brunswick, N.J.: Rutgers University Press.

Agrawal, Arun, and Sanjeev Goyal. 2001. Group size and collective action: Third-party monitoring in common-pool resources. *Comparative Political Studies* 34(1): 63–93.

Agrawal, Arun, and Elinor Ostrom. 2001. Collective action, property rights and decentralization in resource use in India and Nepal. *Politics and Society* 29(4): 485–514.

Agrawal, Arun, and K. Sivaramakrishnan. 2000. Introduction. In *Agrarian Environments: Resources, Representations, and Rule in India*, edited by Arun Agrawal and K. Sivaramakrishnan, 1–22. Durham: Duke University Press.

Agrawal, Arun, and Gautam Yadama. 1997. How do local institutions mediate market and population pressures on resources? Forest panchayats in Kumaon, India. *Development and Change* 28(3): 435–65.

Alaimo, Stacy. 1994. Cyborg and ecofeminist interventions: Challenges for an environmental feminism. *Feminist Studies* 20(1): 133–52.

Albion, R. G. 1926. *Forests and Sea Power: The Timber Problem of the Royal Navy, 1652–1862*. Cambridge, Mass.: Harvard University Press.

Alexander, Paul. 1982. *Sri Lankan Fishermen: Rural Capitalism and Peasant Society*. Canberra: Australian National University.

Allen, E. C. 1904. Letter to the Commissioner, Kumaun Division. Progs. no. 12, serial no. 7, Forest Department File no. 189/E. Lucknow: Uttar Pradesh State Archives.

Alonso, William, and Paul Starr. 1987. Introduction. In *The Politics of Numbers*, edited by William Alonso and Paul Starr, 1–6. New York: Russell Sage Foundation.

Anderson, Benedict. [1983] 1991. *Imagined Communities: Reflections on the Origins and Spread of Nationalism*. New York: Verso.

Anderson, Robert S., and Walter Huber. 1988. *The Hour of the Fox: Tropical Forests, the World Bank, and Indigenous People in Central India*. Seattle: University of Washington Press.

Anonymous. 1912a. Expenditure on forests in India. *Indian Forester* 38(5): 224–28.

——. 1912b. The expenditure on forests in India and its relation to the revenue realized. *Indian Forester* 38(1): 1–17.

——. 1923a. Review of the forest administration report of Madras for the year 1921–22. *Indian Forester* 49(6): 327–30.

——. 1923b. Forest administration report of the United Provinces for the year 1921–22. *Indian Forester* 49(5): 253–57.

——. 1924. Progress report of forest administration in the Punjab. *Indian Forester* 50(6): 263–67.

——. 1926. Review: Forestry in the United Provinces. *Indian Forester* 52(4): 147–50.

——. 1927. Progress report of forest administration in the United Provinces for 1925–26. *Indian Forester* 53(10): 593–96.

——. 1928. Administration report of the Madras Forest Department for 1926–27. *Indian Forester* 54(8): 461–64.

——. 1930. Administration report of the Madras Forest Department for 1928–29. *Indian Forester* 54(8): 355–57.

——. 1931. Forest administration in the United Provinces for the year 1929–30. *Indian Forester* 55(5): 240–41.

Appadurai, Arjun. 1993. Number in the colonial imagination. In *Orientalism and the Postcolonial Predicament: Perspectives on South Asia*, edited by Carol A. Breckenridge and Peter Van der Veer, 314–39. Philadelphia: University of Pennsylvania Press.

——. 1996. *Modernity at Large: Cultural Dimensions of Globalization*. Minneapolis: University of Minnesota Press.

Arizpe, Lourdes, and M. Velazquez. 1994. The social dimensions of population. In *Population and Environment: Rethinking the Debate*, edited by L. Arizpe, M. Priscilla Stone, and D. Major, 15–40. Boulder: Westview.

Arnold, David. 1993. *Colonizing the Body: State Medicine and Epidemic Disease in Nineteenth-Century India*. Delhi: Oxford University Press.

Arnold, David, and Ramachandra Guha, eds. 1995. *Nature, Culture and Imperialism: Essays in the Environmental History of South Asia*. Delhi: Oxford University Press.

Asad, Talal. 1994. Ethnographic representation, statistics, and modern power. *Social Research* 61(1): 55–87.

Ascher, William. 1999. *Why Governments Waste Natural Resources: Policy Failures in Developing Countries*. Baltimore: Johns Hopkins University Press.

Ascher, William, and Robert Healy. 1990. *Natural Resource Policymaking in Developing Countries: Environment, Economic Growth, and Income Distribution*. Durham: Duke University Press.

Atkinson, E. T. [1882] 1973. *The Himalayan Gazetteer*. Vols. 1–3. Delhi: Cosmo Publications.

Ault, Warren Ortman. 1952. *Open Field Farming in Medieval England: The Self-Directing Activities of Village Communities in Medieval England*. Boston: Boston University Press.

Avise, J. 1994. The real message from Biosphere 2. *Conservation Biology* 8(2): 327–29.

Baden-Powell, B. H. 1876. Forest Conservancy in its popular aspect. *Indian Forester* 2(1): 1–16.

——. [1892] 1974. *The Land Systems of British India*. Vol. 2. Oxford: Clarendon Press.

——. 1893. *Forest Law*. London: Bradbury, Agnew, and Co.

Bailey, W. A. 1924. Moribund forests in the United Provinces. *Indian Forester* 50(4): 188–91.

Baistow, K. 1995. Liberation or regulation? Some paradoxes of empowerment. *Critical Social Policy* 42:34–46.

Baland, Jean-Marie, and Jean-Philippe Platteau. 1996. *Halting Degradation of
Natural Resources: Is There a Role for Rural Communities?* Oxford: Clarendon Press.

Bamford, P. W. 1956. *Forests and French Sea Power, 1660–1789.* Toronto: University of Toronto Press.

Barrew, Andrew, Thomas Osborne, and Nikolas Rose, eds. 1996. *Foucault and
Political Reason: Liberalism, Neo-liberalism and Rationalities of Government.*
London: UCL Press.

Barron, A. 1996. The governance of schooling: Genealogies of control and
empowerment in the reform of public education. *Studies in Law, Politics,
and Society* 15:167–204.

Basalla, G. 1967. The spread of Western science. *Science* 156(5, May): 611–22.

Bates, Robert H. 1981. *Markets and States in Tropical Africa.* Los Angeles: University of California Press.

——. 1989. *Beyond the Miracle of the Market: The Political Economy of Agrarian
Development in Kenya.* Cambridge: Cambridge University Press.

——. 2001. *Prosperity and Violence: The Political Economy of Development.* New
York: Norton.

Bates, Robert H., Rui Figueiredo, and Barry R. Weingast. 1998. The politics
of interpretation: Rationality, culture, and transition. *Politics and Society*
26(4): 603–42.

Batten, J. H. 1878. Report on the settlement of district of Garhwal. In *Official
Reports on the Province of Kumaon.* Calcutta: Government Press.

Bauman, Zygmunt. 2000. *Modernity and the Holocaust.* Ithaca: Cornell University Press.

Baumann, Pari. 1995. Decentralizing forest management in India: The case of
van panchayats in Kumaun. Ph.D. thesis, Department of Social and Political Sciences, Darwin College, University of Cambridge.

Bayly, Christopher. 1988. *Indian Society and the Making of the British Empire.*
New York: Cambridge University Press.

Beadon-Bryant, F. 1910. *Review of Forest Administration in British India for the
Year 1907–1908.* Calcutta: Superintendent of Government Printing.

Beddome, R. H. 1878. *Report upon the Nilambur Teak Plantations.* Madras: R.
Hill, Government Press. Letter from Lieutenant-Colonel R. H. Beddome,
conservator of forests, to C. G. Master, Esq., secretary to government,
Revenue Department, 20th April, 1878, .o. 104, p. 61.

Benskin, E. 1924. Correspondence: Moribund forests of the United Provinces. *India Forester* 50(9): 491–93.

Berkes, Fikret, ed. 1989. *Common Property Resources: Ecology and Community-
Based Sustainable Development.* London: Belhaven Press.

Berkes, Fikret, and Cark Folke, eds. 1998. *Linking Social and Ecological Systems: Management Practices and Social Mechanisms for Building Resilience*. Cambridge: Cambridge University Press.

Berlin, Isaiah. 1969. *Four Essays on Liberty*. Oxford: Oxford University Press.

Berry, Sara. 1993. *No Condition Is Permanent: The Social Dynamics of Agrarian Change in Sub-Saharan Africa*. Madison: University of Wisconsin Press.

Blackthorn. 1876. A few notes on "Suggestions regarding forest administration in British Burma." *Indian Forester* 2(2): 190–94.

Blaikie, Piers. 1985. *The Political Economy of Soil Erosion in Developing Countries*. London: Longman.

Blaikie, Piers, and Harold Brookfield. 1987. *Land Degradation and Society*. London: Methuen.

Blascheck, A. D. 1912. A plea for economic forestry. *Indian Forester* 38(3): 115–25

Blomquist, William, and Elinor Ostrom. 1985. Institutional capacity and the resolution of a commons dilemma. *Policies Studies Review* 5(2): 383–93.

Blunt, Peter, and D. Michael Warren. 1996. *Indigenous Organizations and Development*. London: Intermediate Technology.

Bombay Forest Commission. 1887–88. *Report of the Bombay Forest Commission*. Vols. 1–4. Bombay: Government Central Press.

Bose, Sugata, and Ayesha Jalal. 1998. *Modern South Asia: History, Culture, and Political Economy*. London: Routledge.

Bourdieu, Pierre. 1977. *Outline of a Theory of Practice*. Cambridge: Cambridge University Press.

Braddick, Michael J. 2000. *State Formation in Early Modern England, c. 1550–1700*. Cambridge: Cambridge University Press.

Brandis, Dietrich. 1876. *Suggestions Regarding Forest Administration in British Burma*. Calcutta: Office of the Superintendent of Government Printing.

——. 1879. *Review of the Forest Administration in the Several Provinces under the Government of India, 1877–78*. Simla: Government Central Branch Press.

——. 1881a. *Suggestions Regarding Forest Administration in British Burma*. Calcutta: Office of the Superintendent of Government Printing.

——. 1881b. *Review of the Forest Administration in the Several Provinces under the Government of India, 1879–80*. Simla: Government Central Branch Press.

——. [1897] 1994. *Forestry in India: Origins and Early Developments*. New Delhi: Natraj.

——. 1911. *Indian Trees: An Account of Trees, Shrubs, Woody Climbers, Bamboos and Palms Indigenous or Commonly Cultivated in the British Indian Empire*. London: Constable.

Braun, Bruce, and Noel Castree, eds. 1998. *Remaking Reality: Nature at the Millennium*. London: Routledge.

Brenner, Robert. 1977. The origins of capitalist development: A critique of neo-Smithian Marxism. *New Left Review* 104:25–93.

Bromley, Daniel W., ed. 1992. *Making the Commons Work: Theory, Practice, and Policy*. San Francisco: ICS Press.

Brosius, J. Peter. 1997. Prior transcripts, divergent paths: Resistance and acquiescence to logging in Sarawak, East Malaysia. *Comparative Studies in Society and History* 39(3): 468–510.

——. 1999a. Green dots, pink hearts: Displacing politics from the Malaysian rain forest. *American Anthropologist* 101(1): 36–57.

——. 1999b. Analyses and interventions: Anthropological engagements with environmentalism. *Current Anthropology* 40(3): 277–309.

Brow, J. 1996. *Demons and Development: The Struggle for Community in a Sri Lankan Village*. Tucson: University of Arizona Press.

Brown, John Croumbie. 1883. *French Forest Ordinance of 1669 with Historical Sketch of Previous Treatment of Forests in France*. Edinburgh: Oliver and Boyd.

Bryant, Raymond L. 1996. *The Political Ecology of Forestry in Burma*. Honolulu: University of Hawai'i Press.

Bryant, Raymond L., and Sinéad Bailey. 1997. *Third World Political Ecology*. London: Routledge.

Buell, Lawrence. 1995. *The Environmental Imagination: Thoreau, Nature Writing, and the Formation of American Culture*. Cambridge, Mass.: Harvard University Press.

Bunker, S. G. 1985. *Underdeveloping the Amazon: Extraction, Unequal Exchange, and the Failure of the Modern State*. Urbana: University of Illinois Press.

Burchell, Graham, Colin Gordon, and Peter Miller, eds. 1991. *The Foucault Effect: Studies in Governmentality: With Two Lectures by and an Interview with Michel Foucault*. London: Harvester Wheatsheaf.

Butler, Judith. 1989. *Gender Trouble: Feminism and the Subversion of Identity*. New York: Routledge.

——. 1993. *Bodies that Matter*. New York: Routledge.

——. 1997. *The Psychic Life of Power: Theories in Subjection*. Stanford: Stanford University Press.

Calhoun. Craig. 1993. Habitus, field, and capital: The question of historical specificity. In *Bourdieu: Critical Perspectives*, edited by Craig Calhoun, Edward LiPuma, and Moishe Postone, 61–88. Chicago: University of Chicago Press.

Callon, M., and B. Latour. 1981. Unscrewing the big Leviathan: How actors macro-structure reality and how sociology helps them to do so. In *Advances in Social Theory*, edited by K. Knorr-Cetina and A. Cicourel, 280–300. Boston: Routledge.

Campbell, J. S. 1924. Letter to the Secretary of Government of United Provinces. Forest Department File no. 83, Uttar Pradesh State Archives, Lucknow.

Carrier, J. 1987. Marine tenure and conservation in Papua New Guinea: Problems in interpretation. In *The Question of the Commons: The Culture and Ecology of Communal Resources*, edited by Bonnie J. McCay and James M. Acheson, 142–70. Tucson: University of Arizona Press.

Carson, Rachel. 1962. *Silent Spring*. Boston: Houghton Mifflin.

Castree, Noel, and Bruce Braun, eds. 2001. *Social Nature: Theory, Practice, and Politics*. Malden, Mass.: Blackwell Publishers.

Center for Science and Environment (CSE). 1982. *The State of India's Environment: A Citizen's Report*. New Delhi: Center for Science and Environment.

Chakrabarty, Dipesh. 1992. Postcoloniality and the artifice of history: Who speaks for "Indian" pasts? *Representations* 37 (winter): 1–26.

———. 2000a. *Provincializing Europe: Postcolonial Thought and Historical Difference*. Princeton: Princeton University Press.

———. 2000b. Witness to suffering: Domestic cruelty and the birth of the modern subject in Bengal. In *Questions of Modernity*, edited by Timothy Mitchell, 49–86. Minneapolis: University of Minnesota Press.

Chakravarti, Ranabir. 1998. The creation and expansion of settlements and management of hydraulic resources in ancient India. In *Nature and the Orient: The Environmental History of South and Southeast Asia*, edited by Richard Grove, Vinita Damodaran, and Satpal Sangwan, 87–105. Delhi: Oxford University Press.

Champion, H. G. 1919. Observations on some effects of fires in the chir (*Pinus longifolia*) forests of the West Almora division. *Indian Forester* 45:353–63.

———. 1923. The influence of the hand of man on the distribution of forest types in the Kumaon Himalaya. *Indian Forester* 49(3): 116–36.

Champion, Harry, and F. C. Osmaston, eds. 1962. *E. P. Stebbing's "The Forests of India."* Vol. 4: *Being the History from 1925 to 1947 of the Forests Now in Burma, India, and Pakistan*. London: Oxford University Press.

Chatterjee, Partha. 1983. More on modes of power and the peasantry. In *Subaltern Studies: Writings on South Asian History and Society*, edited by Ranajit Guha, 2:311–49. Delhi: Oxford University Press.

———. 1986. *Nationalist Thought and the Colonial World: A Derivative Discourse?* London: Zed.

———. 1993. *The Nation and Its Fragments: Colonial and Postcolonial Histories*. Princeton: Princeton University Press.

Chatterjee, Partha, and Pradeep Jeganathan, eds. 2000. *Subaltern Studies XI: Community, Gender, and Violence*. Delhi: Permanent Black.

Chaturvedi, M. B. 1925. The progress of forestry in the United Provinces. *Indian Forester* 51(4): 357–66.

Chaturvedi, Suprabha, and Vishnu Sahai. 1988. *Institutionalization of Forest Politics: A Study of Jumaon Region in Historical Perspective*. Delhi: Eastern Book Linkers.

Chomitz, Ken. 1995. Roads, land, markets and deforestation: A spatial model of land use in Belize. Paper presented at the first open meeting of the Human Dimensions of Global Environmental Change Community, Duke University, Durham, June 1–3, 1995.

Clay, Jason W. 1988. *Indigenous Peoples and Tropical Forests: Models of Land Use and Management from Latin America*. Cambridge, Mass.: Cultural Survival.

Cline-Cohen, Patricia. 1982. *A Calculating People: The Spread of Numeracy in Early America*. Chicago: University of Chicago Press.

Cockburn, Alex, and J. Ridgeway, eds. 1979. *Political Ecology*. New York: Times Books.

Cohn, Bernard. 1982. The transformation of objects into artifacts, antiquities, and art in nineteenth-century India. In *The Powers of Art: Patronage in Indian Culture*, edited by Barbara Stoler Miller, 301–29. Delhi: Oxford University Press.

——. 1987. The census, social structure, and objectification in South Asia. In *An Anthropologist among the Historians and Other Essays*, 224–54. Delhi: Oxford University Press.

Colburn, F., ed. 1989. *Everyday Forms of Peasant Resistance*. Armonk, N.Y.: M. E. Sharpe.

Colchester, M. 1994. Sustaining the forests: The community-based approach in South and Southeast Asia. *Development and Change* 25(1): 69–100.

Cole, John W., and Eric R. Wolf. [1974] 1999. *The Hidden Frontier: Ecology and Ethnicity in an Alpine Valley*. Berkeley: University of California Press.

Collins, Jane. 1992. Marxism confronts the environment: Labor, ecology and environmental change. In *Understanding Economic Process*, edited by S. Ortiz and S. Lees, 90–112. Lanham, Md.: University Press of America.

Comaroff, Jean, and John Comaroff. 1989. The colonization of consciousness in South Africa. *Economy and Society* 18(3): 267–96.

——. 1991. *Of Revelation and Revolution: Christianity, Colonialism, and Consciousness in South Africa*. Vol. 1. Chicago: University of Chicago Press.

——. 1997. *Of Revelation and Revolution: The Dialectics of Modernity on a South African Frontier*. Vol. 2. Chicago: University of Chicago Press.

Connolly, W. E. 1985. Taylor, Foucault and otherness. *Political Theory* 13(3): 365–76.

Crook, Richard C., and James Manor. 1998. *Democracy and Decentralization in South Asia and West Africa: Participation, Accountability, and Performance*. Cambridge: Cambridge University Press.

Cruikshank, Barbara. 1994. The will to empower: Technologies of citizenship and the war on poverty. *Socialist Review* 23(4): 29–55.

——. 1999. *The Will to Empower: Democratic Citizens and Other Subjects*. Ithaca: Cornell University Press.

Cullen, M. J. 1975. *The Statistical Movement in Early Victorian Britain*. Hassocks, Sussex: Harvester Press.

Dangwal, Dhirendra Datt. 1996. Colonial forestry and agrarian transformation in the U.P. hills, 1815–1947: An agroecological history of the central Himalayas. Ph.D. thesis, Jawaharlal Nehru University, New Delhi.

——. 1997. State, forests, and graziers in the hills of Uttar Pradesh: Impact of colonial forestry on peasants, gujars, and bhotiyas. *Indian Economic and Social History Review* 34(4): 407–35.

——. 1998. Forests, farms and peasants: Agrarian economy and ecological change in the U.P. hills, 1815–1947. *Studies in History*, n.s. 14(2): 349–71.

Darier, Eric, ed. 1999. *Discourses of the Environment*. Malden, Mass.: Blackwell.

de Alessi, Louis. 1980. The economics of property rights: A review of the evidence. *Research in Law and Economics* 2:1–47.

Dean, Mitchell. 1994a. *Critical and Effective Histories*. London: Routledge.

——. 1994b. A social structure of many souls: Moral regulation, government, and self-formation. *Canadian Journal of Sociology* 19(2): 145–68.

——. 1999. *Governmentality: Power and Rule in Modern Society*. London: Sage.

Debord, Guy. [1967] 1994. *The Society of the Spectacle*. New York: Zone Books.

Demeritt, David. 2001. Scientific forest conservation and the statistical picturing of nature's limits in the Progressive-era United States. *Environment and Planning D–Society and Space* 19(4): 431–59.

Dews, Peter. 1984. Power and subjectivity in Foucault. *New Left Review* 144:72–95.

Dhanagare, D. N. 1991. *Peasant Movements in India: 1920–1950*. Delhi: Oxford University Press.

Diamond, Irene, and G. Orenstein. 1988. Ecofeminism: Weaving the worlds together. *Feminist Studies* 14:368–70.

Diamond, Irene, and G. Orenstein, eds. 1990. *Reweaving the World: The Emergence of Ecofeminism*. San Francisco: Sierra Club Books.

Dirks, Nicholas. 1987. *The Hollow Crown: Ethnohistory of an Indian Kingdom*. Cambridge: Cambridge University Press.

Dove, Michael R. 1992. The dialectical history of jungle in Pakistan: An examination of the relationship between nature and culture. *Journal of Anthropological Research* 48(3): 231–53.

Dove, Michael R., and D. M. Kammen. 2001. Vernacular models of development: An analysis of Indonesia under the "New Order." *World Development* 29(4): 619–39.

Drayton, Richard. 2000. *Nature's Government: Science, Imperial Britain, and the "Improvement" of the World*. New Haven: Yale University Press.

Dreyfuss, H., and Paul Rabinow. 1983. *Michel Foucault: Between Structuralism and Hermeneutics*. Chicago: University of Chicago Press.

Durning, A. 1989. *Poverty and the Environment: Reversing the Downward Spiral*. Washington, D.C.: Worldwatch Institute.

Eardly-Wilmot, S. 1906. *Review of Forest Administration in British India, 1903–04*. Calcutta: Office of the Superintendent of Government Printing.

Eardly-Wilmot, S. 1907. *Review of Forest Administration in British India, 1905–06*. Calcutta: Office of the Superintendent of Government Printing.

Eckersley, Robyn. 1992. *Environmentalism and Political Theory: Toward an Ecocentric Approach*. London: UCL Press.

Eckholm, Eric. 1976. *Losing Ground: Environmental Stress and World Food Prospects*. New York: Norton.

Ehrlich, Paul. 1968. *The Population Bomb*. London: Ballantine.

Ehrlich, Paul, and Anne Ehrlich. 1991. *The Population Explosion*. New York: Touchstone.

Ellen, Roy, Peter Parkes, and Alan Bicker, eds. 2000. *Indigenous Environmental Knowledge and Its Transformations: Critical Anthropological Perspectives*. Amsterdam: Harwood Academic Publishers.

Elster, Jon. 1989. *Nuts and Bolts for the Social Sciences*. Cambridge: Cambridge University Press.

——. 1989. *The Cement of Society: A Study of Social Order*. Cambridge: Cambridge University Press.

——. 1999. *Alchemies of the Mind: Rationality and the Emotions*. Cambridge: Cambridge University Press.

Elster, Jon, ed. 1985. *The Multiple Self*. Cambridge: Cambridge University Press.

Elster, Jon, Claus Offe, and Ulrich K. Preuss. 1998. *Institutional Design in Postcommunist Societies: Rebuilding the Ship at Sea*. Cambridge: Cambridge University Press.

Erdosy, George. 1998. Deforestation in pre- and protohistoric South Asia. In *Nature and the Orient: The Environmental History of South and Southeast Asia*, edited by Richard Grove, Vinita Damodaran, and Satpal Sangwan, 51–69. Delhi: Oxford University Press.

Escobar, Arturo. 1995. *Encountering Development: The Making and Unmaking of the Third World*. Princeton: Princeton University Press.

——. 1999. Steps to an antiessentialist political ecology. *Current Anthropology* 40(1): 1–30.

Evans, Peter, Dietrich Rueschemeyer, and Theda Skocpol, eds. 1985. *Bringing the State Back In*. Cambridge: Cambridge University Press.

FAO (Food and Agriculture Organization). 1999. Status and progress in the implementation of national forest programmes: Outcome of an FAO worldwide survey. Rome, FAO. Mimeo.

Fairhead, James, and Melissa Leach. 2000. Desiccation and domination: Science and struggles over environment and development in colonial Guinea. *Journal of African History* 41(1): 35–54.

Farmer, Amy, and Robert H. Bates. 1996. Community versus market. *Comparative Political Studies* 29(4): 379–399.

Farooqui, Amar. 1997. *Colonial Forest Policy in Uttarakhand, 1890–1928*. New Delhi: Kitab Publishing House.

Fearnside, Philip M. 1986. *Human Carrying Capacity of the Brazilian Rainforest.* New York: Columbia University Press.

Fearon, James. 1999. Electoral accountability and the control of politicians: Selecting good type versus sanctioning poor performance. In *Democracy, Accountability, and Representation,* edited by Adam Przeworski and Susan C. Stokes, 55–97. Cambridge: Cambridge University Press.

Ferguson, James. [1990] 1994. *The Anti-politics Machine: "Development," Depoliticization, and Bureaucratic Power in Lesotho.* Minneapolis: University of Minnesota Press.

Ferguson, Kathy. 1993. *The Man Question: Visions of Subjectivity in Feminist Theory.* Berkeley: University of California Press.

Fernandes, Walter, Gita Menon, and Philip Viegas. 1988. *Forests, Environment, and Tribal Economy.* New Delhi: Indian Social Institute.

Fernow, Bernhard E. 1907. *A Brief History of Forestry in Europe, the United States, and Other Countries.* New Haven: Price, Lee and Adkins. Lectures delivered at the Yale Forest School.

Fischer, G. 1993. The population explosion: Where is it leading? *Population and Environment* 15(2): 139–53.

Fisher, Michael H. 1991. *Indirect Rule in India: Residents and the Residency System, 1764–1858.* Delhi: Oxford University Press.

Fisher, W. R. 1885. Forest conservancy at home and abroad. *Indian Forester* 11(12): 586–90.

Forsyth, J. 1872. Highlands of central India: Notes on their forests and wild tribes, natural history, and sports. London: Chapman and Hall.

Foucault, Michel. 1970. *The Order of Things: An Archaeology of the Human Sciences.* New York: Vintage.

———. [1975] 1979. *Discipline and Punish: The Birth of the Prison.* New York: Vintage.

———. 1977. *Power/Knowledge: Selected Interviews and Other Writings, 1972–77,* edited by Colin Gordon, 1–270. New York: Pantheon.

———. [1978] 1991. Governmentality. In *The Foucault Effect: Studies in Governmentality,* edited by Graham Burchell, Colin Gordon, and Peter Miller, 87–104. Chicago: University of Chicago Press.

———. 1978. *The History of Sexuality.* Vol. 1: *An Introduction.* New York: Random House.

———. [1979] 2000. *"Omnes et Singulatim"*: Toward a critique of political reason. In *Michel Foucault: Power,* edited by James D. Faubion, 298–325. New York: Free Press.

———. 1981. The political technology of individuals. In *Technologies of the Self: A Seminar with Michel Foucault,* edited by Luther H. Martin, Huck Gutman, and Patrick H. Hutton, 145–62. Amherst: University of Massachusetts Press.

———. 1982. Afterword: The Subject and Power. In *Michel Foucault: Beyond*

Structuralism and Hermeneutics, edited by Hubert L. Dreyfus and Paul Rabinow, 208–26. Chicago: University of Chicago Press.

——. [1985] 1990. *The History of Sexuality*. Vol. 2: *The Uses of Pleasure*. New York: Vintage.

——. [1986] 1988. *The History of Sexuality*. Vol. 3: *The Care of the Self*. New York: Vintage.

Fox, Jonathan, and David Brown. 1998. *The Struggle for Accountability: The World Bank, NGOs, and Grassroots Movements*. Cambridge, Mass.: MIT Press.

Fox, Richard, and Orin Starn, eds. 1997. *Between Resistance and Revolution: Cultural Politics and Social Protest*. New Brunswick, N.J.: Rutgers University Press.

Fraser, Nancy. 1989. Foucault on modern power: Empirical insights and normative confusions. In *Unruly Practices: Power, Discourse, and Gender in Contemporary Social Theory*, 17–34. Minneapolis: University of Minnesota Press.

Gaard, Greta. 1993. Living interconnections with animals and nature. In *Ecofeminism: Women, Animals, Nature*, edited by Greta Gaard, 1–12. Philadelphia: Temple University Press.

Gadgil, Madhav, and Ramachandra Guha. 1992. *This Fissured Land: An Ecological History of India*. Berkeley: University of California Press.

Gauld, Richard. 2000. Maintaining centralized control in community-based forestry: Policy construction in Philippines. *Development and Change* 31(1): 229–54.

Gaventa, John. 1982. *Power and Powerlessness: Quiescence and Rebellion in an Appalachian Valley*. Urbana: University of Illinois Press.

Geist, Helmut J., and Eric Lambin. 2001. *What Drives Tropical Deforestation? A Meta-analysis of Proximate and Underlying Causes of Deforestation Based on Subnational Case Study Evidence*. LUCC Reports, no. 4. University of Louvain. Louvain-la-Neuve.

Ghai, Dharam. 1993. Conservation, livelihood and democracy: Social dynamics of environmental change in Africa. *Osterreichische Zeitschrift fur Soziologie* 18:56–75.

Gibson, A. J. S. 1920. The coniferous lumber supply and trade of the N.W. Himalayan forests of India. *Indian Forester* 46(7): 355–62.

Gibson, Clark C. 1999. *Politicians and Poachers: The Political Economy of Wildlife Policy in Africa*. New York: Cambridge University Press.

Gibson, Clark C., and Stuart Marks. 1995. Transforming rural hunters into conservationists: An assessment of community-based wildlife management programs in Africa. *World Development* 23(6): 941–95.

Gibson, Clark C., Margaret A. McKean, and Elinor Ostrom, eds. 2000. *People and Forests: Communities, Institutions, and Governance*. Cambridge, Mass.: MIT Press.

Glass, D. V. 1973. *Numbering the People: The Eighteenth Century Population Controversy and the Development of Census and Vital Statistics in Britain*. London: Gordon and Cremonesi.

GOI (Government of India). 1890. *The Indian Forest Act, 1878*. Calcutta: Government Printers.

———. 1930. *Annual Return of Statistics relating to Forest Administration in British India for the year 1928–29*. Calcutta: Central Publications Branch.

———. 1941. *Annual Return of Statistics relating to Forest Administration in British India for the year 1938–39*. Delhi: Manager of Publications.

Gore, Albert. 1992. *Earth in the Balance: Ecology and the Human Spirit*. Boston: Houghton Mifflin.

Goudge, J. E. 1901. Letter to the Commissioner, Kumaun Division. Prog. No. 6, Serial No. 8, Forest Department File No. 197/X-39. Lucknow: Uttar Pradesh State Archives.

GOUP (Government of the United Provinces). 1915. *Annual Progress Report on Forest Administration in the Western, Eastern, and Kumaun Circles of the United Provinces, 1914–15*. Allahabad: Superintendent, Government Press, United Provinces.

———. 1916. *Annual Progress Report of Forest Administration in the United Provinces, 1915–16*. Allahabad: Superintendent, Government Press, United Provinces.

———. 1917. *Annual Progress report on Forest Administration in the United Provinces, 1916–17*. Allahabad: Superintendent, Government Press, United Provinces.

———. 1918. *Annual Progress Report on Forest Administration in the Western, Eastern, and Kumaun Circles of the United Provinces, 1917–18*. Allahabad: Superintendent, Government Press, United Provinces.

———. 1921. *Annual Progress Report of Forest Administration in the United Provinces, 1920–21*. Allahabad: Superintendent, Government Press, United Provinces.

———. 1922. *Annual Progress Report of Forest Administration in the United Provinces, 1921–22*. Allahabad: Superintendent, Government Press, United Provinces.

———. 1925. *Annual Progress Report of Forest Administration in the United Provinces, 1924–25*. Allahabad: Superintendent, Government Press, United Provinces.

———. 1926. *Annual Progress Report of Forest Administration in the United Provinces, 1925–26*. Allahabad: Superintendent, Government Press, United Provinces.

———. 1927. *Annual Progress Report of Forest Administration in the United Provinces, 1926–27*. Allahabad: Superintendent, Government Press, United Provinces.

GOUP. 1931. *Annual Progress Report of Forest Administration in the United Provinces, 1930–31.* Allahabad: Superintendent, Government Press, United Provinces.

——. 1934. *Annual Progress Report of Forest Administration in the United Provinces, 1933–34.* Allahabad: Superintendent, Government Press, United Provinces.

Government of Uttar Pradesh. 1959. *Annual Report on the Working of Forest Panchayats in Kumaun Division, 1957–58.* Allahabad: Superintendent, Printing and Stationery, Uttar Pradesh.

——. 1961. *Annual Report on the Working of Forest Panchayats in Kumaun Division, 1959–60.* Allahabad: Superintendent, Printing and Stationery, Uttar Pradesh.

Grant, B. 1997. Disciplining students: The construction of student subjectivities. *British Journal of Sociology of Education* 18(1): 101–14.

Greenblatt, Stephen. 1999. The touch of the real. In *The Fate of Culture: Geertz and Beyond*, edited by Sherry Ortner, 14–29. Berkeley: University of California Press.

Grendstad, Gunnar, and Dag Wollebaek. 1998. Greener still? An empirical examination of Eckersley's ecocentric approach. *Environment and Behavior* 30(5): 653–75.

Grindle, Merilee Serrill. 2000. Audacious reforms: Institutional invention and democracy in Latin America. Baltimore: Johns Hopkins University Press.

Grove, Richard. 1990. The origins of environmentalism. *Nature* 345:11–14.

——. 1993. Conserving Eden: The East India Companies and their environmental policies on St. Helena, Mauritius, and in western India, 1660–1854. *Comparative Studies in Society and History* 35:318–51.

——. 1995. *Green Imperialism: Colonial Expansion, Tropical Island Edens, and the Origins of Environmentalism, 1600–1860.* Cambridge: Cambridge University Press.

——. 1998a. Indigenous knowledge and the significance of south-west India for Portuguese and Dutch constructions of tropical nature. In *Nature and the Orient: The Environmental History of South and Southeast Asia*, edited by Richard Grove, Vinita Damodaran, and Satpal Sangwan, 187–209. Delhi: Oxford University Press.

——. 1998b. *Ecology, Climate, and Empire: The Indian Legacy in Global Environmental History, 1400–1940.* Delhi: Oxford University Press.

Grove, Richard, Vinita Damodaran, and Satpal Sangwan. 1998. Introduction. In *Nature and the Orient: The Environmental History of South and Southeast Asia*, edited by Richard Grove, Vinita Damodaran, and Satpal Sangwan, 1–26. Delhi: Oxford University Press.

Grover, Ruhi. 1997. Rhythms of the timber trade: Forests in the Himalayan Punjab, 1850–1925. Ph.D. diss., University of Virginia.

Guha, Ramachandra. 1983. Forestry in British and post-British India: An historical analysis. *Economic and Political Weekly* 17:1882–96.

——. 1990. An early environmental debate: The making of the 1878 Forest Act. *Indian Economic and Social History Review* 27(1): 65–84.

——. 2000. Epilogue (1998): The Afterlifes of Chipko. In *The Unquiet Woods: Ecological Change and Peasant Resistance in the Himalaya*. 2d ed. Berkeley: University of California Press.

——. [1989] 2000. *The Unquiet Woods: Ecological Change and Peasant Resistance in the Himalaya*. New Delhi: Oxford University Press.

Guha, Ranajit. 1982a. On some aspects of the historiography of colonial India. In *Subaltern Studies I: Writings on South Asian History and Society*, 1–8. Delhi: Oxford University Press.

——. 1983. *Elementary Aspects of Peasant Insurgency in Colonial India*. Delhi: Oxford University Press.

——. 1997. *Dominance without Hegemony: History and Power in Colonial India*. Cambridge, Mass.: Harvard University Press.

Guha, Ranajit, ed. 1982b. *Subaltern Studies I: Writings on South Asian History and Society*. Delhi: Oxford University Press.

Guha, Ranajit, and Gayatri Chakravorty Spivak, eds. 1988. *Selected Subaltern Studies*. New York: Oxford University Press.

Guha, Sumit. 1999. *Environment and Ethnicity in India, 1200–1991*. Cambridge: Cambridge University Press.

——. 2000. Economic rents and natural resources: Commons and conflicts in premodern India. In *Agrarian Environments: Resources, Representations, and Rule in India*, edited by Arun Agrawal and K. Sivaramakrishnan, 132–46. Durham: Duke University Press.

Gupta, Akhil. 1995. Blurred boundaries: The discourse of corruption, the culture of politics, and the imagined state. *American Ethnologist* 22(2): 375–402.

——. 1998. *Postcolonial Developments: Agriculture in the Making of Modern India*. Durham: Duke University Press.

Gurung, Barun. 1992. Towards sustainable development: A case in the eastern Himalayas. *Futures* 24:907–916.

Gururani, Shubhra. 1996. Fuel, fodder, and forests: Politics of forest use and abuse in Uttarakhand Himalaya, India. Ph.D. diss., Syracuse University.

Hacking, Ian. [1981] 1991. How should we do the history of statistics? In *The Foucault Effect: Studies in Governmentality*, edited by Graham Burchell, Colin Gordon, and Peter Miller, 181–96. Chicago: University of Chicago Press.

——. 1986. Making up people. In *Reconstructing Individualism: Autonomy, Individuality, and the Self in Western Thought*, edited by T. C. Heller, M. Sosna, and D. E. Welbery, 222–36. Stanford: Stanford University Press.

Hacking, Ian. 1990. *The Taming of Chance*. Cambridge: Cambridge University Press.

——. 1999. *The Social Construction of What?* Cambridge, Mass.: Harvard University Press.

Hall, J., ed. 1986. *States in History*. Oxford: Basil Blackwell.

Hannah, Matthew G. 2000. *Governmentality and the Mastery of Territory in Nineteenth-Century America*. Cambridge: Cambridge University Press.

Haraway, Donna. 1989. *Primate Visions: Gender, Race, and Nature in the World of Modern Science*. New York: Routledge.

——. 1991. *Simians, Cyborgs, and Women: The Reinvention of Nature*. New York: Routledge.

Hardiman, David, ed. 1992. *Peasant Resistance in India, 1858–1914*. Delhi: Oxford University Press.

Hardin, Garrett. 1968. The tragedy of the commons. *Science* 162:1243–48.

——. 1978. Political requirements for preserving our common heritage. In *Wildlife and America*, edited by H. P. Bokaw, 310–17. Washington, D.C.: Council on Environmental Quality.

——. 1993. *Living within Limits*. New York: Oxford University Press.

Harrison, Robert Pogue. 1992. *Forests: The Shadow of Civilization*. Chicago: University of Chicago Press.

Haynes, Douglas, and Gyan Prakash, eds. 1992. *Contesting Power: Resistance and Social Relations in South Asia*. Delhi: Oxford University Press.

Heilbroner, Robert. 1974. *An Inquiry into the Human Prospect*. New York: Norton.

Herbert-Cheshire, L. 2000. Contemporary strategies for rural community development in Australia: A governmentality perspective. *Journal of Rural Studies* 16(2): 203–15.

Herbst, Jeffrey. 2000. *States and Power in Africa: Comparative Lessons in Authority and Control*. Princeton: Princeton University Press.

Heske, Franz. 1938. *German Forestry*. New Haven: Yale University Press.

Hill, Kevin A. 1996. Zimbabwe's Wildlife Utilization Programs: Grassroots Democracy or an Extension of State Power? *African Studies Review* 39(1): 103–21.

Hirsch, Eric. 1999. Colonial units and ritual units: Historical transformations of persons and horizons in highland Papua. *Comparative Studies in Society and History* 41: 805–28.

Hobsbawm, Eric, and Terence Ranger. 1983. *The Invention of Tradition*. Cambridge: Cambridge University Press.

Holdren, C. 1992. Population alarm. *Science* 255:1358.

Holmstrom, Bengt. 1982. Moral hazard in teams. *Bell Journal of Economics* 13(2): 324–40.

Hopwood, A. G., and P. Miller, eds. 1994. *Accounting as Social and Institutional Practice*. Cambridge: Cambridge University Press.

Howard, S. H. 1937. *Forest Pocketbook*. Allahabad: Superintendent, Printing and Stationery.

Hughes, D. M. 2001. Cadastral politics: The making of community-based resource management in Zimbabwe and Mozambique. *Development and Change* 32(4): 741–68.

Inden, Ronald. 1990. *Imagining India*. Oxford: Basil Blackwell.

Irvine, J., et al., eds. 1979. *Demystifying Social Statistics*. London: Pluto Press.

Ives, J., and D. Pitt, eds. 1988. *Deforestation: Social Dynamics in Watersheds and Mountain Ecosystems*. London: Routledge.

Jackson, Cecile. 1993a. Doing what comes naturally? Women and environment in development. *World Development* 21(12): 1947–63.

———. 1993b. Environmentalisms and gender interests in the third world. *Development and Change* 24:649–77.

———. 1993c. Women/nature or gender/history: A critique of ecofeminist "development." *Journal of Peasant Studies* 20(3): 389–418.

Jackson, H. 1911. Letter from H. Jackson, Conservator of forests, Eastern circle to the Chief Secretary to the Government, United Provinces, no. 40-L.G., dated 11th July, 1910, Forest Department Progs no. 76, Serial no. 3, May 1911, p. 113.

Jackson, H. 1985. A survey of agriculture in Kumaon. Mirtola Ashram, Almora. Mimeo.

Jardine, Nicholas. 1991. *The Scenes of Inquiry: On the Reality of Questions in the Sciences*. Oxford: Clarendon.

Jaroff, L. 1993. Happy birthday, double helix. *Time* 141(11): 56–59.

Jessop, Bob. 1990. *State Theory: Putting the Capitalist State in Its Place*. Cambridge: Polity Press.

Jewett, Sara. 2000. Mothering earth? Gender and environmental protection in the Jharkhand, India. *Journal of Peasant Studies* 27(2): 94–131.

Jodha, Narpat. 1990. Rural common property resources: Contributions and crisis. Foundation day lecture, May 16, 1990, New Delhi, Society for the Promotion of Wasteland Development. Mimeo.

Kaplan, M., and J. D. Kelly. 1994. Rethinking resistance: Dialogics of disaffection in colonial Fiji. *American Ethnologist* 21(1): 123–51.

Katten, Michael. 1999. Manufacturing village identity and its village: The view from nineteenth-century Andhra. *Modern Asian Studies* 33(1): 87–120.

Kessinger, Tom. 1974. *Vilayatpur, 1848–1968: Social and Economic Change in a North Indian Village*. Berkeley: University of California Press.

KFGC (Kumaon Forest Grievances Committee). 1921. *Report of the Forest Grievances Committee for Kumaon*. Government of Uttar Pradesh.

Khare, Arvind, Madhu Sarin, N. C. Saxena, Subhabrata Patil, Seem Bathla, Farhad Vania, and M. Satyanarayana. 2000. *Joint Forest Management: Pol-*

icy, Practice, and Prospects. London: International Institute for Environment and Development.

King, Katie. 1994. *Theory in Its Feminist Travels: Conversations in U.S. Women's Movements*. Bloomington: Indiana University Press.

Kirkwood, A. 1893. *Papers and Reports upon Forestry, Forest Schools, Forest Administration, and Management in Europe, America, and the British Possessions*. Toronto: Warwick and Sons.

Klooster, D. 2000. Community forestry and tree theft in Mexico: Resistance or complicity in conservation? *Development and Change* 31(1): 281–305.

——. 2002. Toward adaptive community forest management: Integrating local forest knowledge with scientific forestry. *Economic Geography* 78(1): 43–70.

Knight, Jack. 1992. *Institutions and Social Conflict*. Cambridge: Cambridge University Press.

Kripke, Saul. 1972. *Wittgenstein on Rules and Private Language*. Cambridge, Mass.: Harvard University Press.

Kula, Witold. 1986. *Measures and Men*. Translated by R. Szerter. Princeton: Princeton University Press.

Kumaon, Deepak. 1990. The evolution of colonial science: Natural history and the East India Company. In *Imperialism and the Natural World*, edited by J. M. McKenzie, 51–66. Manchester: Manchester University Press.

Kumar, Deepak. 1980. Patterns of colonial science. *Indian Journal of History of Science* 15(1): 105–13.

Kumar, Neeraj. 2001. All is not green with JFM in India. *Forests, Trees, and People Newsletter* 42:46–50.

Lal, Makhan. 1985. Iron tools, forest clearance, and urbanization in the Gangetic plains. *Man and Environment* 10:83–90.

Lam, Wai Fung. 1998. *Governing Irrigation Systems in Nepal: Institutions, Infrastructure, and Collective Action*. Oakland: Institute for Contemporary Studies Press.

Latour, Bruno. 1986. The powers of association. In *Power, Action, and Belief*, edited by J. Law, 32–57. London: Routledge and Kegan Paul.

——. 1987. *Science in Action*. Cambridge, Mass.: Harvard University Press.

——. 1999. *Pandora's Hope: Essays on the Reality of Science Studies*. Cambridge, Mass.: Harvard University Press.

Lazarsfeld, P. F. 1961. Notes on the history of quantification in sociology: Trends, sources, and problems. *Isis* 52:277–333.

Leach, Melissa. 1994. *Rainforest Relations: Gender and Resource Use among the Mende of Gola, Sierra Leone*. Washington, D.C.: Smithsonian Institution Press.

Lesch, John E. 1990. Systematics and the geometrical spirit. In *The Quantifying Spirit in the 18th Century*, edited by Tore Frängsmyr, J. L. Heilbron, and Robin E. Rider, 73–111. Berkeley: University of California Press.

Lévi-Strauss, C. 1985. *The View from Afar*. Chicago: University of Chicago Press.

Lewis, Oscar. 1958. *Village Life in Northern India*. Urbana: University of Illinois Press.

Li, Tania M. 1996. Images of community: Discourse and strategy in property relations. *Development and Change* 27(3): 501–27.

———. 1999. Compromising power: Development, culture, and rule in Indonesia. *Cultural Anthropology* 14(3): 195–322.

———. 2000. Articulating indigenous identity in Indonesia: Resource politics and the tribal slot. *Comparative Studies in Society and History*. 42(1): 149–79.

———. 2001. Masyarakat adat, difference, and the limits of recognition in Indonesia's forest zone. *Modern Asian Studies* 35:645–76.

Lichbach, Mark I. 1998. Contending theories of contentious politics and the structure-action problem of social order. *Annual Review of Political Science* 1:401–24.

Lindqvist, Svante. 1990. Labs in the woods: The quantification of technology during the late Enlightenment. In *The Quantifying Spirit in the 18th Century*, edited by Tore Frängsmyr, J. L. Heilbron, and Robin E. Rider, 291–314. Berkeley: University of California Press.

Lloyd, D., and P. Thomas. 1998. *Culture and the State*. New York: Routledge.

Lloyd, William, and Alexander Gerard. 1840. *Narrative of a journey from Caunpoor to the Boorendo Pass in the Himalaya Mountains via Gwalior, Agra, Delhi, and Sirhind*, edited by George Lloyd. London: J. Madden.

Low, B., and J. Heinen. 1993. Population, resources and environment: Implications of human behavioral ecology for conservation. *Population and Environment* 15(1): 7–41.

Lowood, Henry E. 1990. The calculating forester: Quantification, cameral science, and the emergence of scientific forestry in Germany. In *The Quantifying Spirit in the 18th Century*, edited by Tore Frängsmyr, J. L. Heilbron, and Robin E. Rider, 315–43. Berkeley: University of California Press.

Ludden, David. 1993. Orientalist empiricism: Transformations of colonial knowledge. In *Orientalism and the Postcolonial Predicament: Perspectives on South Asia*, edited by Carol A. Breckenridge and Peter Van der Veer, 250–78. Philadelphia: University of Pennsylvania Press.

Ludden, David, ed. 2001. *Reading Subaltern Studies: Critical History, Contested Meaning, and the Globalisation of South Asia*. Delhi: Permanent Black.

Lugard, Frederick D. 1926. *The Dual Mandate in British Tropical Africa*. 3d ed. Edinburgh: William Blackwood and Sons.

Luke, Timothy W. 1995. On environmentality, geo-power, and eco-knowledge in the discourse of contemporary environmentalism. *Cultural Critique* 31(fall): 57–81.

Luke, Timothy W. 1997. *Ecocritique: Contesting the Politics of Nature, Economy, and Culture*. Minneapolis: University of Minnesota Press.

Lukes, Steven. 1974. *Power: A Radical View*. London: Macmillan.

Lynch, Owen, and Kirk Talbott. 1995. *Balancing Acts: Community-Based Forest Management and National Law in Asia and the Pacific*. Baltimore: World Resources Institute.

MacLeod, Roy, 1975. Scientific advice for British India: Imperial perceptions and administrative goals, 1896–1923. *Modern Asian Studies* 9(3): 343–84.

MacLeod, Roy and M. Lewis, eds. 1988. *Disease, Medicine, and Empire: Perspectives in Western Medicine and the Experience of European Expansion*. London: Routledge.

MacLeod, Roy, and Deepak Kumar. 1995. *Technology and the Raj: Western Technology and Technology Transfers in India, 1700–1947*. New Delhi: Sage.

Malthus, T. [1798, 1803] 1960. *On Population (First Essay on Population, 1798, and Second Essay on Population, 1803)*. New York: Random House.

Mamdani, Mahmood. 1996. *Citizen and Subject: Contemporary Africa and the Legacy of Late Colonialism*. Princeton: Princeton University Press

——. 2001. *When Victims Become Killers: Colonialism, Nativism, and the Genocide in Rwanda*. Princeton: Princeton University Press.

Mann, Michael. 1988. *States, War and Capitalism*. Oxford: Blackwell.

Marriott, McKim, ed. 1955. *Village India: Studies in the Little Community*. Chicago: University of Chicago Press.

Mawdsley, Elizabeth. 1998. After Chipko: From environment to region in Uttaranchal. *Journal of Peasant Studies* 25(4): 36–54.

McCarthy, Nancy, Brent Swallow, Michael Kirk, and Peter Hazell. 1999. *Property Rights, Risk, and Livestock Development in Africa*. Washington, D.C.: ILRI and IFPRI.

McCay, Bonnie J., and James Acheson, eds. 1987. *The Question of the Commons: The Culture and Ecology of Communal Resources*. Tucson: University of Arizona Press.

McKean, Margaret. 1992. Success on the commons: A comparative examination of institutions for common property resource management. *Journal of Theoretical Politics* 4(3): 247–81.

——. 2000. Common property: What is it, what is it good for, and what makes it work? In *People and Forests: Communities, Institutions, and Governance*, edited by Clark C. Gibson, Margaret A. McKean, and Elinor Ostrom, 27–55. Cambridge, Mass.: MIT Press.

McMahon, Martha. 1997. From the ground up: Ecofeminism and ecological economics. *Ecological Economics* 20:163–73.

McNeely, J. A., ed. 1995. *Expanding Partnerships in Conservation*. Washington, D.C.: Island Press.

Meadows, D., D. Meadows, and J. Randers. 1992. *Beyond the Limits: Confront-

ing Global Collapse, Envisioning a Sustainable Future. Post Mills: Chelsea Green.

Meadows, D., J. Randers, and W. Behrens. 1972. *The Limits to Growth*. New York: Universe Books.

Meffe, G., A. Ehrlich, and D. Ehrenfeld. 1993. Human population control: The missing agenda. *Conservation Biology* 7(1): 1–3.

Mellor, Mary. 1992. *Breaking the Boundaries: Toward a Feminist Green Socialism*. London: Virago Press.

———. 1997. *Feminism and Ecology*. New York: New York University Press.

Merchant, Carolyn. 1980. *The Death of Nature: Women, Ecology, and the Scientific Revolution*. San Francisco: Harper and Row.

———. 1990. Ecofeminism and ecofeminist theory. In *Reweaving the World: The Emergence of Ecofeminism*, edited by Irene Diamond and G. Orenstein, 102–28. San Francisco: Sierra Club Books.

Mies, Maria, and Vandana Shiva. 1993. *Ecofeminism*. Atlantic Highlands, N.J.: Zed.

Migdal, Joel. 1988. *Strong Societies, Weak States: State-Society Relations and State Capacity in the Third World*. Princeton: Princeton University Press.

Migdal, Joel, Atul Kohli, and Vivian Shue, eds. 1994. *State Power and Social Forces: Domination and Transformation in the Third World*. Cambridge: Cambridge University Press.

Miller, Gary. 1992. *Managerial Dilemmas: The Political Economy of Hierarchy*. New York: Cambridge University Press.

Miller, Peter. 1994. Accounting and objectivity: The invention of calculating selves and calculable spaces. In *Rethinking Objectivity*, edited by Alan Megill, 239–64. Durham: Duke University Press.

Miller, Peter, and T. O'Leary. 1987. Accounting and the construction of the governable person. *Accounting, Organizations and Society* 12:235–65.

Miller, Peter, and Nikolas Rose. 1990. Governing economic life. *Economy and Society* 19(1): 1–31.

Miller, Shawn William. 2000. *Fruitlesss Trees: Portuguese Conservation and Brazil's Colonial Timber*. Stanford: Stanford University Press.

Miller, Toby. 1993. *The Well-Tempered Self: Citizenship, Culture, and the Postmodern Subject*. Baltimore: Johns Hopkins University Press.

Misra, V. N. 1989. Stone age India: An ecological perspective. *Man and Environment* 14:17–64.

Mitchell, Timothy. 1990. Everyday metaphors of power. *Theory and Society* 19:545–77.

———. 1991. The limits of the state: Beyond statist approaches and their critics. *American Political Science Review* 85:77–96.

Mitchell, Timothy, ed. 2000. *Questions of Modernity*. Minneaoplis: University of Minnesota Press.

Mobbs, E. C. 1929. The cowdung problem or village forests. *Indian Forester* 55(9): 469–80.

Molesworth, G. L. 1880. Durability of Indian Railway sleepers and the rules for marking them. *Indian Forester* 6(2): 97–99.

Monier-Williams, Monier. 1891. *Brahmanism and Hinduism, 4th ed*. London: J. Murray.

Moore, Barrington, Jr. 1966. *Social Origins of Dictatorship and Democracy*. Boston: Beacon Press.

———. 1978. *Injustice: The Social Basis of Obedience and Revolt*. White Plains: M. E. Sharpe.

Moore, Donald S. 1996a. Marxism, culture and political ecology: Environmental struggles in Zimbabwe's eastern highlands. In *Liberation Ecologies: Environment, Development, Social Movements*, edited by Richard Peet and Michael Watts, 125–47. New York: Routledge.

———. 1996b. A river runs through it: Environmental history and the politics of community in Zimbabwe's eastern highlands. *Journal of Southern African Studies* 24(2): 377–403.

———. 1998. Subaltern struggles and the politics of place: Remapping resistance in Zimbabwe's eastern highlands. *Cultural Anthropology* 13(3): 344–81.

———. 1999. The crucible of cultural politics: Reworking "development" in Zimbabwe's eastern highlands. *American Ethnologist* 26(3): 654–89.

Munro, William A. 1998. *The Moral Economy of the State: Conservation, Community Development, and State-Making in Zimbabwe*. Athens: Ohio University, Center for International Studies.

Murali, Atluri. 1996. Whose trees? Forest practices and local communities in Andhra, 1600–1922. In *Nature, Culture, Imperialism: Essays on the Environmental History of South Asia*, edited by David Arnold and Ramachandra Guha, 86–122. Delhi: Oxford University Press.

Myers, Norman. 1991. The world's forests and human populations: The environmental interconnections. In *Resources, Environment, and Population: Present Knowledge, Future Options*, edited by K. Davis and M. Bernstam, 237–51. New York: Oxford University Press.

Nagarkoti, Diwan. 1997. Uttarakhand ki lattha panchayaten. *Pahar* 10:268–80.

National Research Council. 1986. *Proceedings of the Conference on Common Property Resource Management*. Washington, D.C.: National Academy Press.

———. 1999. *Our Common Journey: A Transition toward Sustainability*. Washington, D.C.: National Academy Press.

Nef, John U. 1966. *The Rise of the British Coal Industry*. 2 vols. Hamden, Conn.: Archon Books.

Nelson, Diane. 1997. Crucifixion stories, the 1869 Caste War of Chiapas, and negative consciousness: A disruptive subaltern study. *American Ethnologist* 24(2): 331–54.

Ness, G., W. Drake, and S. Brechin, eds. 1993. *Population-Environment Dynamics: Ideas and Observations*. Ann Arbor: University of Michigan Press.

Netting, Robert M. 1981. *Balancing on an Alp*. Cambridge: Cambridge University Press.

Neumann, Roderick P. 1992. The political ecology of wildlife conservation in the Mount Meru area of northeast Tanzania. *Land Degradation and Rehabilitation* 3(2): 85–98.

——. 1998. *Imposing Wilderness: Struggles over Livelihood and Nature Preservation in Africa*. Berkeley: University of California Press.

Neumann, R. P., and R. A. Schroeder, eds. 1995. *Manifest Ecological Destinies: Local Rights and Global Environmental Agendas*. Special issue. *Antipode* 27:321–428.

Neville, H. R. 1904. *Nainital: A Gazetteer*. Allahabad: Superintendent, Printing and Stationery, United Provinces.

Nicholson, Jack. 1991. Solar electricity. In *Heaven Is under Our Feet*, edited by D. Henley and D. Marsh, 28–32. New York: Berkeley Books.

North, Douglass C. 1990. *Institutions, Institutional Change and Economic Performance*. Cambridge: Cambridge University Press.

Nugent, David. 1994. Building the state, making the nation: The bases and limits of state centralization in "modern" Peru. *American Anthropologist* 96(2): 333–69.

O'Malley, P. 1992. Risk, power, and crime prevention. *Economy and Society* 21(3): 252–75.

Ophuls, William. 1973. Leviathan or Oblivion. In *Toward a Steady State Economy*, edited by H. E. Daly, 43–58. San Francisco: Freeman.

——. 1977. *Ecology and the Politics of Scarcity: Prologue to a Political Theory of the Steady State*. San Francisco: W. H. Freeman.

Osmaston, Arthur Edward. 1921. *Working Plan for the North Garhwal Forest Division, 1921–22 to 1930–31*. Allahabad: Government Press.

——. 1927. *A Forest Flora for Kumaon*. Allahabad: Superintendent, Government Press, United Provinces.

Ostrom, Elinor. 1990. *Governing the Commons: The Evolution of Institutions for Collective Action*. New York: Cambridge University Press.

——. 1999. Self Governance and Forest Resources. Occasional Paper No. 20, Center for International Forestry Research, Bogor, Indonesia.

Ostrom, Elinor, Roy Gardner, and James Walker. 1994. *Rules, Games, and Common-Pool Resources*. Ann Arbor: University of Michigan Press.

Ostrom, Elinor, Larry Schroeder, and Susan Wynne. 1993. *Institutional Incentives and Sustainable Development: Infrastructure Policies in Perspective*. Boulder: Westview.

Ostrom, Elinor, James Walker, and Roy Gardner. 1992. Covenants with and without a sword: Self-Governance is possible. *American Political Science Review* 86(2): 404–17.

Pande, Badri Datta. 1993. History of Kumaun (English version of Kumaun ka Itihas). Almora: Shyam Prakashan.

Pant, G. 1922. *The Forest Problem in Kumaon*. Nainital: Gyanodaya Prakashan.

Pant, S. D. 1935. *The Social Economy of the Himalayans*. London: George Allen and Unwin.

Parnwell, Michael, and Raymond Bryant. 1996. *Environmental Change in Southeast Asia: People, Politics, and Sustainable Development*. London: Routledge.

Pasquino, P. 1991. Theatrum politicum: The genealogy of capital–police and the state of prosperity. In *The Foucault Effect: Studies in Governmentality*, edited by G. Burchell, Colin Gordon, and Peter Miller, 105–18. Chicago: University of Chicago Press.

Pathak, Shekhar. 1991. The *begar* abolition movements in British Kumaun. *Indian Economic and Social History Review* 28(3): 261–79.

Patriarca, Silvana. 1996. *Numbers and Nationhood: Writing Statistics in Nineteenth-Century Italy*. Cambridge: Cambridge University Press.

Patton, Paul. 1989. Taylor and Foucault on power and freedom. *Political Studies* 37:260–76.

Pauw, E. K. 1896. Report on the Tenth Settlement of the Garhwal District. Allahabad: Superintendent, Government Press.

Pavlich, G. 1996. The power of community mediation: Government and formation of self-identity. *Law and Society Review* 30(4): 707–33.

Pearson, G. 1869. *Report on the Forests of Garhwal and Kumaon*. Allahabad: Government Press.

Pearson, J. R. 1926. Note on the history of proposals for management of village waste lands. In *FD File #83/1909*. Lucknow: Uttar Pradesh State Archives.

Pearson, Ralph S. 1925. *The Development of India's Forest Resources*. Calcutta: Government of India Central Publications Branch.

Peet, Richard, and Michael Watts. 1996. Toward a theory of liberation ecology. In *Liberation Ecologies: Environment, Development, Social Movements*, edited by Richard Peet and Michael Watts, 260–69. London: Routledge.

Peluso, Nancy L. 1992. *Rich Forests, Poor People: Resource Control and Resistance in Java*. Berkeley: University of California Press.

Peluso, Nancy L., and Peter Vandergeest. 2002. Genealogies of the political forest and customary rights in Indonesia. *Journal of Asian Studies* 60(3): 761–812.

Peluso, Nancy L., and Michael Watts, eds. 2001. *Violent Environments*. Ithaca: Cornell University Press.

Peters, Pauline E. 1994. *Dividing the Commons: Politics, Policy, and Culture in Botswana*. Charlottesville: University Press of Virginia.

Philip, Kavita Sara. 1996. The role of science in colonial discourses and practices of modernity: Anthropology, forestry, and the construction of nature's resources in Madras forests, 1858–1930. Ph.D. diss., Cornell University.

Philp, Mark. 1983. Foucault on power: A problem in radical translation. *Political Theory* 11(1): 29–52.

Pickering, Andrew. 1995. *The Mangle of Practice: Time, Agency, and Science.* Chicago: University of Chicago Press.

Pigg, Stacy L. 1992. Inventing social categories through place: Social representations and development in Nepal. *Comparative Studies in Society and History* 34(3): 491–518.

Pimental, D., R. Harman, M. Pacenza, J. Pecarsky, and M. Pimental. 1994. Natural resources and an optimal human population. *Population and Environment* 15(5): 347–69.

Pinkerton, Evelyn, ed. 1989. *Cooperative Management of Local Fisheries: New Directions for Improved Management and Community Development.* Vancouver: University of British Columbia Press.

Pinkerton, Evelyn, and Martin Weinstein. 1995. *Sustainability through Community-Based Management.* Vancouver: David Suzuki Foundation.

Poffenberger, Mark, ed. 1990. *Keepers of the Forest: Land Management Alternatives in Southeast Asia.* West Hartford, Conn.: Kumarian.

Poffenberger, Mark, and Betsy McGean, eds. 1996. *Village Voices, Forest Choices: Joint Forest Management in India.* Delhi: Oxford University Press.

Porter, Theodore. 1986. *The Rise of Statistical Thinking, 1820–1900.* Princeton: Princeton University Press.

Porter, Theodore. 1996. *Trust in Numbers: The Invention of Objectivity.* Princeton: Princeton University Press.

Prakash, Gyan. 1990. Writing post-orientalist histories of the third world: Perspectives from Indian historiography. *Comparative Studies in Society and History* 32(2): 383–408.

———. 1999. *Another Reason: Science and the Imagination of Modern India.* Princeton: Princeton University Press.

———. 2000. Body politic in colonial India. In *Questions of Modernity*, edited by Timothy Mitchell, 189–222. Minneapolis: University of Minnesota Press.

Prasad, Archana. 1994. Forests and subsistence in colonial India: A study of the Central Provinces, 1830–1945. Ph.D. diss., Jawaharlal Nehru University.

Pratap, Ajay. 2000. *The Hoe and the Axe: An Ethnohistory of Cultivation in Eastern India.* New Delhi: Oxford University Press.

Procacci, G. 1991. Social economy and the government of poverty. In *The Foucault Effect: Studies in Governmentality*, edited by G. Burchell, Colin Gordon, and Peter Miller, 151–68. Chicago: University of Chicago Press.

Putnam, Robert D. 1993. *Making Democracy Work: Civic Traditions in Modern Italy*. Princeton: Princeton University Press.

Rabinow, Paul. 1984. Introduction. In *The Foucault Reader*, 1–29. New York: Pantheon.

Raffles, Hugh. 1998. On the Nature of the Amazon: The Politics and imagination of an anthropogenic landscape. Ph.D. diss., Yale University.

———. 2002. Intimate Knowledge. *International Social Science Journal* 54(3): 325–34.

Rajan, Ravi S. 1998a. Imperial environmentalism or environmental imperialism: European forestry, colonial foresters, and the agendas of forest management in British India, 1800–1900. In *Nature and the Orient: The Environmental History of South and Southeast Asia*, 324–71. Delhi: Oxford University Press.

———. 1998b. Foresters and the politics of colonial agroecology: The case of shifting cultivation and soil erosion, 1920–1950. *Studies in History*, n.s. 14(2): 217–36.

Rangan, Haripriya. 1997. Property vs. control: The state and forest management in the Indian Himalaya. *Development and Change* 28(1): 71–94.

———. 2000. *Of Myths and Movements: Rewriting Chipko into Himalayan History*. London: Verso.

Rangarajan, L. N. 1987. *Kautilya: The Arthashastra*. New Delhi: Penguin.

Rangarajan, Mahesh. 1994. Imperial agendas and India's forests: The early history of Indian forestry, 1800–1878. *Indian Economic and Social History Review* 31(2): 147–67.

———. 1996. *Fencing the Forest: Conservation and Ecological Change in India's Central Provinces, 1860–1914*. Delhi: Oxford University Press.

Rawat, A. S. 1983. *Garhwal Himalayas, a Historical Survey: The Political and Administrative History of Garhwal, 1815–1947*. Delhi: Eastern Books

———. 1989. *History of Garhwal, 1358–1947: An Erstwhile Kingdom in the Himalayas*. New Delhi: Indus Publishing.

———. 1991. History of forest management, conflicts, and their impact in the central Himalaya. In *The History of Forestry in India*, 280–325. New Delhi: Indus Publishing.

Redclift, Michael. 1984. *Development and the Environmental Crisis: Red or Green Alternatives*. London: Methuen.

Redford, Kent H., and Christine Padoch, eds. 1992. *Conservation of Neotropical Forests: Working from Traditional Resource Use*. New York: Columbia University Press.

Regmi, Mahesh C. 1999. *Imperial Gorkha: An Account of Gorkhali Rule in Kumaun, 1791–1915*. Delhi: Adroit Publishers.

Repetto, Robert, and Malcolm Gillis, eds. 1988. *Public Policies and the Misuse of Forest Resources*. Cambridge: Cambridge University Press.

Ribbentrop, Berthold. 1896. *Review of Forest Administration in British India for the Year 1894–95*. Calcutta: Superintendent of Government Printing.

———. 1899. *Forestry in British India*. Calcutta: Government of India Central Printing Office.

———. 1900. *Forestry in British India*. Calcutta: Office of the Superintendent of Government Printing.

Ribot, Jesse. 1995. From Exclusion to participation: Turning Senegal's forestry policy around. *World Development* 23(9): 1587–99.

———. 1999. Decentralisation, participation and accountability in Sahelian forestry: Legal instruments of political-administrative control. *Africa* 69(1): 23–65.

Robertson, F. 1936. *Our Forests*. Lucknow: Government Press.

———. 1942. Fifteen years of forest administration in the United Provinces: A short retrospect, I–II. *Indian Forester* 68(2–3): 55–64, 115–22.

Rocheleau, Diane. 1995. Gender and biodiversity: A feminist political ecology perspective. *IDS Bulletin* 26(1): 9–16.

Rodger, A. 1930. Quinquennial Review of Forest Administration in British India, ending 31st March, 1929. Calcutta: Central Publications Branch.

Rorty, Richard. 1984. Habermas and Lyotard on postmodernity. *Praxis International* 4(1): 32–44.

Rose, Nikolas S. [1989] 1999. *Governing the Soul: The Shaping of the Private Self*. London: Free Association Books.

———. 1996. *Inventing Our Selves: Psychology, Power, and Personhood*. London: Cambridge University Press.

———. 1999. *Powers of Freedom: Reframing Political Thought*. Cambridge: Cambridge University Press.

Saberwal, Vasant. 1999. *Pastoral Politics: Shepherds, Bureaucrats, and Conservation in the Himachal Himalaya*. Delhi: Oxford University Press.

Sahlins, Peter. 1994. *Forest Rites: The War of the Demoiselles in Nineteenth-Century France*. Cambridge, Mass.: Harvard University Press.

Said, Edward. 1978. *Orientalism: Western Representations of the Orient*. London: Routledge.

Salleh, Ariel. 1984. Deeper than deep ecology: The eco-feminist connection. *Environmental Ethics* 6: 340–45.

———. 1997. *Ecofeminism as Politics: Nature, Marx, and the Postmodern*. London: Zed.

Sangwan, Satpal. 1994. Re-ordering the earth: The emergence of geology as a scientific discipline in colonial India. *Indian Economic and Social History Review* 33(3): 291–308.

———. 1998. From gentlemen amateurs to professionals: Reassessing the natural science tradition in colonial India, 1780–1840. In *Nature and the Orient: The Environmental History of South and Southeast Asia*, edited by Richard

Grove, Vinita Damodaran, and Satpal Sangwan, 211–36. Delhi: Oxford University Press.

Sarin, Madhu. 2002. From right holders to "beneficiaries": Community forest management, van panchayats, and village forest joint management in Uttarakhand. Mimeo.

Satz, Debra, and John Ferejohn. 1994. Rational choice and social theory. *Journal of Philosophy* 91:71–87.

Schelling, Thomas. 1980. *The Strategy of Conflict*. Cambridge, Mass.: Harvard University Press.

Schlager, Edella, and Elinor Ostrom. 1992. Property rights regimes and natural resources: A conceptual analysis. *Land Economics* 68(3): 249–62.

Schlich, W. 1906. *Schlich's Manual of Forestry*. Vol. 1: *Forest Policy in the British Empire*. 3d ed. London: Bradbury, Agnew.

——. 1910. *Schlich's Manual of Forestry*. Vol. 2: *Silviculture*. 4th ed. London: Bradbury, Agnew.

——. 1911. *Schlich's Manual of Forestry*. Vol. 3: *Forest Management*. 4th ed. London: Bradbury, Agnew.

Scott, David. 1995. Colonial governmentality. *Social Text* 43 (fall): 191–220.

——. 1999. *Refashioning Futures: Criticism after Postcoloniality*. Princeton: Princeton University Press.

Scott, James C. 1976. *The Moral Economy of the Peasant: Rebellion and Subsistence in Southeast Asia*. New Haven: Yale University Press.

——. 1985. *Weapons of the Weak: Everyday Forms of Peasant Resistance*. New Haven: Yale University Press.

——. 1990. *Domination and the Arts of Resistance: Hidden Transcripts*. New Haven: Yale University Press.

——. 1998. *Seeing Like a State: How Certain Schemes to Improve the Human Condition Have Failed*. New Haven: Yale University Press.

Seager, Joni. 1993. *Earth Follies: Coming to Feminist Terms with the Global Environmental Crisis*. New York: Routledge.

Sengupta, Nirmal. 1991. *Managing Common Property: Irrigation in India and the Philippines*. London: Sage.

Shapiro, Ian. 1989. Gross concepts in political argument. *Political Theory* 17(1): 51–76.

Shaw, Norman. 1914. *Chinese Forest Trees and Timber Supply*. London: T. Fisher Unwin.

Sherring, C. A. 1904. Letter from C. A. Sherring to the Commissioner, Kumaun Division. Progs. no. 2, Serial no. 1, Forest Department File no. 164, Lucknow, Uttar Pradesh State Archives.

Shiva, Vandana. 1988. *Staying Alive: Women, Ecology, and Survival*. London: Zed.

——. 1994. *Close to Home: Women Reconnect Ecology, Health, and Development Worldwide*. Philadelphia: New Society Publications.

Shrivastava, Aseem. 1996. Property rights, deforestation, and community forest management in the Himalayas: An analysis of forest policy in British Kumaun, 1815–1949. Ph.D. diss., Department of Economics, University of Massachusetts, Amherst.

Siddiqi, Majid Hayat. 1978. *Agrarian Unrest in North India: The United Provinces, 1918–22*. New Delhi: Vikas.

Simpson, John. 1900. *The New Forestry: Continental System Adapted to British Woodlands and Game Preservation*. Sheffield: Pawson and Brailsford.

Singh, Chetan. 1996. Forests, pastoralists, and agrarian society in Mughal India. In *Nature, Culture, Imperialism: Essays on the Environmental History of South Asia*, edited by David Arnold and Ramachandra Guha, 21–48. Delhi: Oxford University Press.

——. 1998. *Natural Premises: Ecology and Peasant Life in the Western Himalaya, 1800–1950*. Delhi: Oxford University Press.

Sivaramakrishnan, K. 1995. Imagining the past in present politics: Colonialism and forestry in India. *Comparative Studies in Society and History* 37(1): 3–40.

——. 1996. Forests, politics and governance in Bengal, 1794–1994. Ph.D. diss., Department of Anthropology, Yale University.

——. 1999. *Modern Forests: Statemaking and Environmental Change in Colonial Eastern India*. Stanford: Stanford University Press.

——. 2000. Crafting the public sphere in the forests of West Bengal: Democracy, development and political action. *American Ethnologist* 27(2): 431–61.

Skaria, Ajay. 1999. *Hybrid Histories: Forests, Frontiers, and Wildness in Western India*. Delhi: Oxford University Press.

Smythies, E. A. 1911. Some aspects of fire protection in chir forests. *Indian Forester* 37(1–2): 54–62.

——. 1914. *The Resin Industry in Kumaon*. Forest Bulletin no. 26. Allahabad: Superintendent, Printing and Stationery, United Provinces.

——. 1917. Possibilities of development in the Himalayan coniferous forests. *Indian Forester* 43(4): 165–72.

Somanathan, E. 1990. "Public forests and private interests." New Delhi, Indian Statistical Institute. Mimeo.

——. 1991. Deforestation, property rights, and incentives in Central Himalaya. *Economic and Political Weekly* 26:PE37–46.

Spivak, Gayatri C. 1988. *In Other Worlds*. New York: Routledge.

Srinivas, M. N., ed. 1960. *India's Villages*. Bombay: Asia Publishing House.

Starhawk. 1982. *Dreaming the Dark*. Boston: Beacon Press.

Stebbing, Edward P. 1916. *British Forestry: Its Present Position and Outlook after the War*. London: John Murray.

——. 1919. *Commercial Forestry in Britain: Its Decline and Revival*. London: John Murray.

Stebbing, E. P. 1922–26. *The Forests of India*. Vols. 1–3. London: The Bodley Head.

——. 1928. *The Forestry Question in Great Britain*. London: The Bodley Head.

——. 1937. *The Forests of West Africa and the Sahara: A Study of Modern Conditions*. London: W. and R. Chambers.

Stein, Burton. 1980. *Peasant State and Society in Medieval South India*. Delhi: Oxford University Press.

——. 1985. State formation and economy reconsidered. *Modern Asian Studies* 19(3): 387–413.

——. 1989. *Thomas Munro: The Origins of the Colonial State and His Vision of Empire*. Delhi: Oxford University Press.

Steinmetz, George, ed. 1999. *State/Culture: State Formation after the Cultural Turn*. Ithaca: Cornell University Press.

Stocks, Anthony. 1987. Resource management in an Amazon Varzea Lake ecosystem: The Cocamilla case. In *The Question of the Commons: The Culture and Ecology of Communal Resources*, edited by Bonnie J. McCay and James M. Acheson, 108–20. Tucson: University of Arizona Press.

Stoler, Ann Laura. 1989. Rethinking colonial categories: European communities and the boundaries of rule. *Comparative Studies in Society and History* 31:134–61.

——. 1995. *Race and the Education of Desire: Foucault's* History of Sexuality *and the Colonial Order of Things*. Durham: Duke University Press.

Stowell, V. A. 1911. Letter no. 2443/XXV–7 from V. A. Stowell, Deputy Commissioner of Garhwal to Commissioner Kumaon Division, dated 18 June 1910. Progs. no. 80, Forest Department Serial no. 6.

——. 1916. Note on proposals for village forests under the Nayabad Act. Forest Department File no. 195, Uttar Pradesh State Archives, Lucknow.

Street, Richard. 1998. Tattered shirts and ragged pants: Accommodation, protest, and the coarse culture of California wheat harvesters and threshers, 1866–1900. *Pacific Historical Review* 67(4): 573–608.

Sturgeon, Noël. 1997. *Ecofeminist Natures: Race, Gender, Feminist Theory and Political Action*. New York: Routledge.

——. 1999. Ecofeminist appropriations and transnational environmentalisms. *Identities* 6(2–3): 255–79.

Sundar, Nandini. 1997. *Subalterns and Sovereigns: An Anthropological History of Bastar*. Delhi: Oxford University Press.

——. 2000. Unpacking the "joint" in joint forest management. *Development and Change* 31(1): 255–60.

Tang, Shui Yan. 1992. *Institutions and Collective Action: Self Governance in Irrigation Systems*. San Francisco: ICS Press.

Taylor, Charles. 1955. *Philosophy and the Human Sciences*. Cambridge: Cambridge University Press.

Taylor, Charles. 1979. *Hegel and Modern Society*. Cambridge: Cambridge University Press.

———. 1984. Foucault on freedom and truth. *Political Theory* 12:152–83.

———. 1993. To follow a rule . . . In *Bourdieu: Critical Perspectives*, edited by Craig Calhoun, Edward LiPuma, and Moishe Postone, 45–60. Chicago: University of Chicago Press.

Taylor, Michael. 1982. *Community, Anarchy, and Liberty*. Cambridge: Cambridge University Press.

Thirsk, Joan. 1966. The origins of common fields. *Past and Present* 33(April): 142–47.

Thomas, N. 1994. *Colonialism's Culture: Anthropology, Travel, and Government*. Cambridge: Cambridge University Press.

Tiffen, Mary, Michael Mortimore, and Francis Gichuki. 1994. *More People, Less Erosion: Environmental Recovery in Kenya*. Chichester, N.Y.: Wiley.

Tolia, R. S. 1994. *British Kumaun-Garhwal: An Administrative History of a Non-regulation Hill Province, Gardner and Traill Years, 1815–1835*. Almora: Shree Almora Book Depot.

Traill, G. W. [1828] 1992. Statistical Sketch of Kamaon. *Asiatic Researches* 16:137–234.

Trevor, C. G., and E. A. Smythies. 1923. *Practical Forest Management: A Handbook with Special Reference to the United Provinces of Agra and Oudh*. Allahabad: Government Press, United Provinces.

Troup, Robert Scott. 1909. *Indian Woods and Their Uses*. Calcutta: Superintendent of Government Printing.

Tucker, Richard. 1983. The British colonial system and the forests of the western Himalayas, 1815–1914. In *Global Deforestation and the Nineteenth Century World Economy*, edited by Richard Tucker and J. F. Richards, 146–66. Durham: Duke University Press.

———. 1988a. The depletion of India's forests under British imperialism: Planters, foresters, and peasants in Assam and Kerala. In *The Ends of the Earth: Perspectives on Modern Environmental History*, edited by Donald Worster, 118–40. Cambridge: Cambridge University Press.

———. 1988b. The British Empire and India's forest resources: The timberlands of Assam and Kumaon, 1914–1950. In *World Deforestation in the Twentieth Century*, edited by John F. Richards and Richard P. Tucker, 91–111. Durham: Duke University Press.

Tully, James. 1988. Governing conduct. In *Conscience and Casuistry in Early Modern Europe*, edited by Edmund Leites, 12–71. New York: Cambridge University Press.

Turner, J. E. C. 1924. Urgent and confidential letter to Mr. Canning, 22.2.1924. Forest Department File no. 271, Uttar Pradesh State Archives, Lucknow.

Turner, J. E. C. 1929. West Almora Division, U.P. *Indian Forester* 55(11): 578–86.

Unwin, A. Harold. 1920. *West African Forests and Forestry*. London: T. Fisher Unwin.

UPFD (United Provinces Forest Department). 1925. *Annual Progress Report of Forest Administration in the United Provinces, 1923–24*. Allahabad: Superintendent, Printing and Stationery, United Provinces.

———. 1930. *Annual Progress Report of Forest Administration in the United Provinces, 1929–30*. Allahabad: Superintendent, Government Press, United Provinces.

———. 1940. *Annual Progress Report of Forest Administration in the United Provinces, 1939–40*. Allahabad: Superintendent, Government Press, United Provinces.

———. 1961. *The Centenary of Forest Administration in Uttar Pradesh, 1861–1961*. Lucknow: Uttar Pradesh Forest Department.

Uttar Pradesh Government. 1931. Forest Council Rules of 1931. Lucknow: Government Printers.

———. 1959. *Annual Progress Report of the Forest Councils in Kumaon and Garhwal, 1958–59*. Lucknow: Government Printers.

———. 1976. Forest Council Rules of 1976. Lucknow: Government Printers.

Vandergeest, Peter. 1996. Property rights in protected areas: Obstacles to community involvement as a solution in Thailand. *Environmental Conservation* 23(3): 259–68.

Varughese G., and E. Ostrom. 2001. The contested role of heterogeneity in collective action: Some evidence from community forestry in Nepal. *World Development* 29 (5): 747–65.

Verma, L. R., and T. Partap. 1992. The experiences of an area based development strategy in Himachal Pradesh, India. In *Sustainable Mountain Agriculture*, edited by N. S. Jodha, M. Banskota, and Tej Partap, 2:76–101. New Delhi: Oxford.

Veyne, Paul. 1997. Foucault revolutionises history. In *Foucault and His Interlocutors*, edited by A. Davidson, 146–82. Chicago: University of Chicago Press.

Vico, Giambattista. 1968. *The New Science*. Ithaca: Cornell University Press.

Vigne, G. T. 1842. *Travels in Kashmir, Ladak, Iskardo, the Countries adjoining the Mountain Course of the Indus and the Himalaya, North of the Panjab, with Map*. Vol. 2. London: Henry Colburn.

Voelcker, J. A. 1897. *Report on the Improvement of Indian Agriculture*. Calcutta: Government of India Press.

Wade, Robert. [1988] 1994. *Village Republics: Economic Conditions for Collective Action in South India*. Oakland: ICS Press.

Walton, H. G. 1911. *Almora: A Gazetteer*. Allahabad: Superintendent, Printing and Stationery, United Provinces.

Warren, Kay. 1987. Feminism and ecology: Making connections. *Environmental Ethics* 9(1): 3–20.

Warren, Kay. 1994. *Ecological Feminism*. New York: Routledge.

Warren, Kay, ed. 1997. *Ecofeminism: Women, Nature, Culture*. Indianapolis: Indiana University Press.

Washbrook, David. 1988. Progress and problems: South Asian economic and social history, c. 1720–1860. *Modern Asian Studies* 22(1): 57–96.

Washbrook, David, and Rosalind O'Hanlon. 1992. After orientalism: Culture, criticism, and politics in the third world. *Comparative Studies in Society and History* 34(1): 141–67.

Watts, Michael. 1983. *Silent Violence: Food, Famine, and Peasantry in Northern Nigeria*. Berkeley: University of California Press.

Weaver, James H., Michael T. Rock, and Kenneth Kusterer. 1997. *Achieving Broad-Based Sustainable Development: Governance, Environment, and Growth with Equity*. West Hartford, Conn.: Kumarian.

Weber, Max. 1958. *The Religion of India: The Sociology of Hinduism and Buddhism*. Translated and edited by H. H. Gerth and Don Martindale. Glencoe, Ill.: Free Press.

Weber, T. W. 1866. *Report on the Operation of Forest Survey in Kumaon and Garhwal*. Allahabad: Government Press.

Weber, Thomas. 1988. *Hugging the Trees: The Story of the Chipko Movement*. New Delhi: Viking.

Wells, David, and Tony Lynch. 2000. *The Political Ecologist*. Aldershot: Ashgate.

Wells, Michael. 1998. Institutions and incentives for biodiversity conservation. *Biodiversity and Conservation* 7(6): 815–35.

Western, David, and R. Michael Wright, eds. 1994. *Natural Connections: Perspectives in Community-Based Conservation*. Washington, D.C.: Island Press.

White, Richard. 1995. *The Organic Machine: The Remaking of the Columbia River*. New York: Hill and Wang.

Whited, Tamara L. 2000. *Forests and Peasant Politics in Modern France*. New Haven: Yale University Press.

Wildavsky, A. 1987. Choosing preferences by constructing institutions: A cultural theory of preference formation. *American Political Science Review* 81(1): 3–21.

———. 1994. Why self-interest means less outside of a social context: Cultural contributions to a theory of rational choices. *Journal of Theoretical Politics* 6(2): 131–59.

Willis, Paul. 1981. *Learning to Labor: How Working Class Kids Get Working Class Jobs*. New York: Columbia University Press.

Wilson, E. O. 1992. *The Diversity of Life*. New York: Norton.

Wittgenstein, Ludwig. 1973. *Philosophical Investigations*. Translated by G. E. M. Anscombe. Oxford: Basil Blackwell.

Wolf, E. 1972. Ownership and political ecology. *Anthropological Quarterly* 45(3): 201–5.

Wolverkamp, Paul, ed. 1999. *Forests for the Future: Local Strategies for Forest Protection, Economic Welfare and Social Justice*. London: Zed.

Woolsey, Theodore S., Jr. 1917. *French Forests and Forestry: Tunisia, Algeria, Corsica*. New York: Wiley.

——. 1920. *Studies in French Forestry*. New York: Wiley.

Worby, Eric. 2000. "Discipline without oppression": Sequence, timing, and marginality in Southern Rhodesia's post-war development regime. *Journal of African History* 41(1): 101–25.

Wyndham, P. 1916. Letter to the Lt. Governor of U.P., Forest Department File no. 40, Uttar Pradesh State Archives, Lucknow.

Yelling, J. A. 1977. *Common Field and Enclosure in England, 1450–1850*. London: MacMillan.

Young, Kenneth R. 1994. Roads and the environmental degradation of tropical montane forests. *Conservation Biology* 8(4): 972–76.

Zolberg, Aristide. 1972. Moments of madness. *Politics and Society* 2(2): 183–207.

Index

Note: *f* with page number indicates figure; *t* indicates table.

Acheson, James, 131
action at a distance, 193, 194
adjudication systems, 132–33, 136*t*, 154–55, 157
Agrawal, Arun, 241 n.6, 261 n.11, 266 nn.5, 6
Agrawal, Bina, 203–4, 212, 268 n.17
agriculture, forests and, 79, 252 n.29
Airadeo, India, forest fires set in, 4
Albion, R. G., 238 n.4
all and each, government of, 218, 270 n.27
Allen, E. C., 174
allocation systems: community-based, 134, 135*t*, 136–40, 158, 159, 160; Forest Council Rules of 1931 and, 132; of forest department, 140, 157; political ecology theory on, 210; sanctioning systems and, 147
Almora, India, 1, 79
Anaimalai teak forests, management of, 45–46
Anderson, Benedict, 167–68, 171, 263 nn.2, 4
Anderson, T., 46
annual reviews of forest administration, 49–50; on burden of regulations and grazing fees, 255 n.2–3; on effects of Kumaon Forest Grievances Committee recommendations, 83–84, 246 n.32, 253 nn.38, 39; on forest fires set by villagers, 175, 252 n.32;
antiessentialist theory, 211
Appadurai, Arjun, 172, 241 n.4

Arizpe, Lourdes, 129
Asad, Talal, 241 n.4
assigned mutual monitoring, 145, 187, 188*t*
Atkinson, E. T., 68, 248 n.3
auctions, allocation by, 138, 139–40

Baden-Powell, B. H., 75, 79, 85, 251 n.21, 252 n.30
Bahuguna, Sundar Lal, 231 n.1
Bailey, Sinéad, 211, 268 n.14
Bailey, W. A., 249 n.8
Baland, Jean-Marie, 131, 207, 261 n.9
Bates, Robert H., 264 n.11
Batten (Kumaon commissioner), 68–69, 256 n.7
Baumann, Zygmunt, 111, 229, 256 nn.3, 5
Beckmann, Johann Gottlieb, 243 n.15
Beddome, R. H., 53
Bengal, forest management in, 46
Berlin, Isaiah, 220, 272 n.38
Berry, Sara, 254 n.4
Bhagartola council, Almora district: allocation regulations in, 139–40; data on, 263 n.21; forest statistics of, 149*t*; sanctioning system of, 148–53
Bhatt, Chandi Prasad, 231 n.1
biodiversity, decline in, 230
biological traits, feminism and, 269 n.18
biopower: Escobar on, 273 n.45; Foucault on, 97, 217, 270 nn.25, 26
Blaikie, Piers, 210
Bombay: breaches of forest regulations in, 105; clashes in forest and revenue departments of, 76; For-

and, 35–38; effects on Indian subjects of, 172; Forest Council Rules
of 1931 and, 102; forest making
and, 4, 19, 28, 29–30, 60–61, 231
n.5; labels for forest management
by, 240 n.10; lattha panchayats as
regulations model for, 108–9; political rationality for colonialized
and, 254 n.6; rise of illegality and,
232 nn.9, 11; in South Asia, 232
n.13; villagers' relationship to forests and, 179–80. *See also* British
East India Company; forest departments; statistics and statistical knowledge
Comaroff, Jean and John, 234 n.25
commerce: conservation and, 102,
255 n.1; forest department control
of, 116, 118–19, 259 n.22. *See also*
revenue from forests
commercial uses for forests: important species for, 68–69, 70, 250
n.11. *See also* revenue from forests
Commission on Survey, Bombay,
India, 27
common property, 203; as choice
for discussion, 266–67 n.8; constraints in theories on, 214; environmentality and, 23;
environmental politics and theories on, 205–8, 266 n.4, 267 n.10;
political ecology and, 208–9. *See
also* community-based conservation
communities: politics involved in,
218–19; subject formation and
politicization of, 171; use of term,
89. *See also* regulatory communities
community-based conservation:
colonial rule and, 256 n.5; commercial exploitation of forests
and, 102, 255 n.1; forest making
and, 102, 255 n.1; growth of, 127; inter-

national examples of, 14–15, 202,
235 n.31, 235–36 n.32, 260 n.3, 266
n.5; private management *versus*,
237 n.48; regulatory tools for, 133–
34; research on effectiveness of,
267 n.9; separation between state
and, 66–67; studies on local use
and control for, 131–32, 260 n.1,
261 nn.9, 11; village-based forest
council variations and, 130–31.
See also lattha panchayats; regulatory communities; village-level
forest councils
conduct of conduct: definition of,
269 n.23; in domain of environment, 222; sources of, 7. *See also*
governmentality; governmentality, Foucauldian
Congress Party, Civil Disobedience
movement, 253 n.35
Connolly, Mr., Nilumbur teak plantations and, 53
Connolly, W. E., 272 n.35
consciousness, changes in forest
subjectivities and, 234 n.25
conservation of forests, 250 n.14;
Brandis's assessment of, in
Burma, 40; Brandis's rules for,
43–45; classifications and quantification of, 39; creation of
village-level forest councils and,
83–84, 253 n.40; exclusionist, 72–
73, 104–5, 205; knowledge forms
and, 227–28; for profitable exploitation, 102, 255 n.1; Ramsay's
measures of, 69–70; social practices and, 207. *See also*
community-based conservation;
fire protection; Kumaon forests;
working plans
constitution of unified categories,
feminism and, 269 n.18
construction of forests, definition
of, 34. *See also* making of forests

ARUN AGRAWAL is associate professor

in the School of Natural Resources and

the Environment at the University of

Michigan, Ann Arbor.

Library of Congress Cataloging-in-Publication Data

Agrawal, Arun, 1962–
Environmentality : technologies of government
and the making of subjects / Arun Agrawal.
p. cm. — (New ecologies for the twenty-first
century)
Includes bibliographical references and index.
ISBN 0-8223-3480-1 (cloth : alk. paper)
ISBN 0-8223-3492-5 (pbk. : alk. paper)
1. Environmental policy—Decision making. 2.
Environmental management—Decision making.
3. Environmental policy—India. 4. Environmental management—India. 5. Forest conservation—Kumaun Himalaya. I. Title. II. Series.
GE170.A26 2005
333.72—dc22   2004025374